Hooray for Subtraction Facts!

by
Becky Daniel

illustrated by Judy Hierstein

Math Facts Wheel by Kenneth Holland

Cover by Judy Hierstein

Copyright © Good Apple, Inc., 1990

Printing No. 987654321

ISBN No. 0-86653-518-7

GOOD APPLE, INC.
299 JEFFERSON ROAD, P.O. BOX 480
PARSIPPANY, NJ 07054-0480

Table of Contents

GA1134

To the Teacher

Mastering the subtraction facts doesn't have to be a boring tedious process, not if you use *Hooray for Subtraction Facts!* The puzzles, mazes, codes, magic tricks, games, learning aids and math facts wheel found in this book were carefully designed to teach the basic subtraction facts while entertaining and delighting young mathematicians. Every page is a different activity format so children never get bored.

It is highly recommended that you begin this book by giving each student the three-page subtraction facts test found on pages 63, 64, and 65. Allow exactly five minutes for the students to answer as many of the subtraction problems as they can. Record the initial scores. As the children are learning the facts, retest them weekly using the same three-page test and recording the scores each time. You may want the children to graph their own progress. (See graph on page 66.) The test can also be used as a subtraction facts aid. Simply cut along the solid lines to create mini flash cards so children can review the subtraction facts with which they are experiencing difficulty. Each child should also make his/her own subtraction facts wheel. See reproducible patterns on pages 71 and 72.

The many activity sheets found in *Hooray for Subtraction Facts!* were carefully designed and sequenced so that each page adds new facts and reviews those already learned. With the exception of several activities in the back of the book, no regrouping is necessary. If children master the basic facts, mathematics is enjoyable; however, if they do not master those facts, mathematics will soon become difficult and frustrating. Don't let one single student leave your classroom this year not knowing the subtraction facts. It's fun and it's easy to sharpen math skills with the colossal collection of ideas found herein.

GA1134

Using the Math Facts Wheel

The self-teaching math wheel is a fantastic teaching tool for grades two, three and four. Students using this simple device learn the basic math facts without consuming valuable class teaching time. The teacher is freed from forcing exercises upon students because students actually enjoy the math wheel. Learning subtraction facts becomes fun. The embarrassment that sometimes accompanies the use of flash cards is eliminated. To use the wheel, the student first gives his answer to himself, then advances the wheel slightly and presto! the correct answer appears! If he was correct, seeing the correct answer will reinforce the student's memory or confidentially correct his error. Moving through the full range of basic subtraction facts on the wheel, the student rapidly achieves readiness to perform the basic math functions.

This math tool has demonstrated its usefulness with special students who are having learning difficulties. Slow learners require continuing reinforcement and correction. The math wheel meets these needs. Continued practice day after day helps these students achieve using the wheel without the teacher's constant supervision or assistance.

Assembly of the Math Facts Wheel

1. Punch out the two die cut wheels.
2. Punch out the small answer box sections from the smaller (top) wheel.
3. Using a metal brad fastener, fasten the wheels together with the smaller wheel on top. Spread the brad tips behind the large (bottom) wheel.

Assembly of Teacher-Produced Wheels

1. Reproduce a copy of each of the components of the wheel for each of your students.
2. Glue these copies to oaktag or poster board.
3. Cut out, using heavy scissors or sharp knife. (Do not allow students to use knives.)
4. Cut out answer windows with X-acto knife. It is important to stay within the windows when cutting.
5. Assemble as with steps used on original wheel (see instructions above).

How to Use the Math Facts Wheel

1. Hold the assembled wheel in both hands arranging the perimeter numbers of both wheels so that they match up.
2. For example, for the subtraction wheel, first move the top wheel so that the 1 printed on its edge is directly below the 2 on the edge of the bottom wheel. The student should then think "2 − 1 = 1." Now move the top wheel slightly clockwise. The answer 1 should appear in the answer window to reinforce or correct the student.
3. Continue moving the top wheel until its number 1 is directly below the 3 on the bottom wheel. Give your answer. Move the top wheel slightly clockwise. The answer 2 will appear.
4. Continue in this fashion with the 1's going up to 24 − 1. Then proceed with the 2's, applying it to each number on the bottom wheel, etc.

GA1134

When the minuend (number subtracting from) and the subtrahend (number being subtracted) are the same, the difference is always ZERO. Example: If you have two cookies and you eat two cookies, you have ZERO cookies left. Complete the subtraction problems found below.

1 - 1 =   **100 - 100 =**

5 - 5 =   **1000 - 1000 =**

When you subtract ZERO (subtrahend) from another number (minuend), the difference is the minuend. Example: If you have three apples and I take away ZERO apples, you still have three apples. Complete the subtraction problems found below with ZERO as the subtrahend.

6 - 0 =   **10 - 0 =**

9 - 0 =   **50 - 0 =**

Write two subtraction problems with equal minuends and subtrahends.

1.   **2.**

Write two subtraction problems with subtrahends of ZERO.

1.   **2.**

Bonus: When is it possible to borrow one dollar from your friend and not owe him/her a dollar?

    1   GA1134

Miss One

Cross out apples to show each subtraction problem found below.

2 - 1 =

1 - 1 =

1 - 0 =

2 - 2 =

Work your way through the maze found below, subtracting each number you pass. There are many paths through the maze, but only one path has a difference of ONE! Can you find what path?

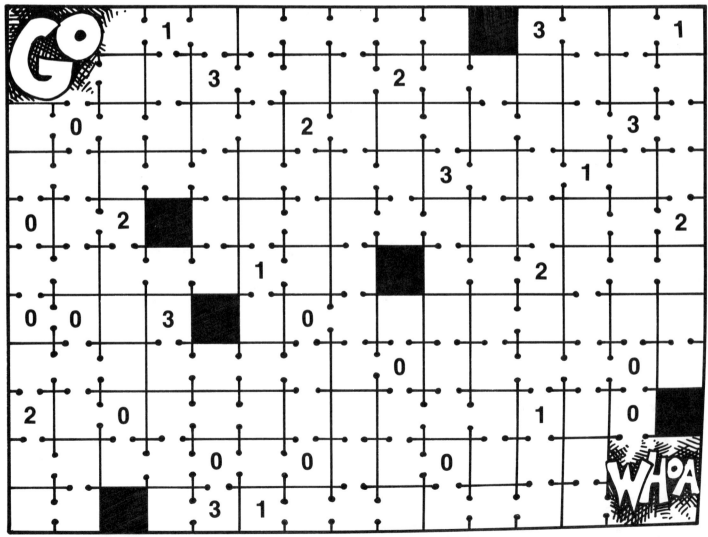

Bonus: Draw your own subtraction maze.

GA1134

Mrs. Two

Use the number line to help you solve the subtraction problems found below. Begin by finding the minuend on the number line. Move to the left two places to discover the differences. Then color all the spaces with an even difference RED. Color the spaces with an odd difference BLUE. (The even numbers have been circled on the number line.)

⓪ 1 ② 3 ④ 5 ⑥ 7 ⑧ 9 ⑩ 11 ⑫ 13 ⑭

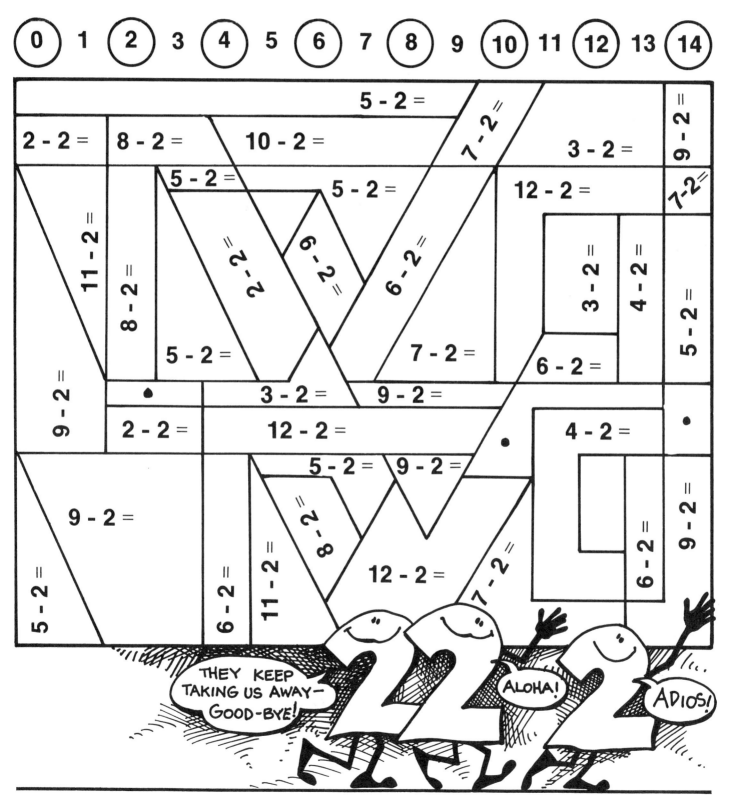

3

Dr. Three

Subtract the number of dots in the right half from the number of dots in the left half of each domino. Write each problem and then solve it. The first one has been done for you.

1. $3 - 0 = 3$

2.

3.

4.

5.

6.

7.

Bonus: Draw ten different dominoes that have three dots more on the left than on the right side.

GA1134

Professor Four

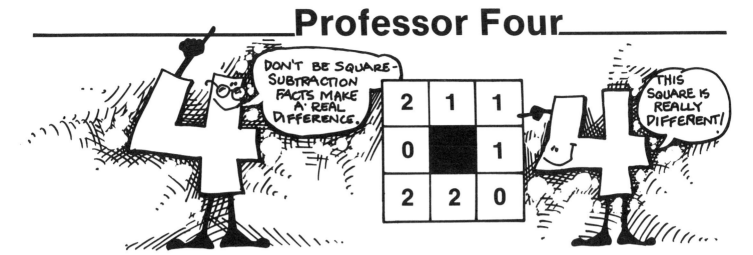

Complete each magic square. The last number in each row and each column must be the difference between the first and second number in that row or column. See the example that has been completed for you.

1.
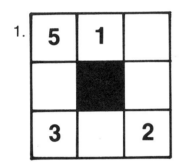

2.
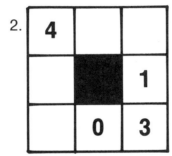

3.

4.
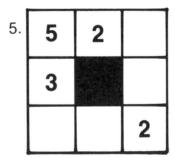

5.

6.

Bonus: Make up three of your own subtraction magic squares.

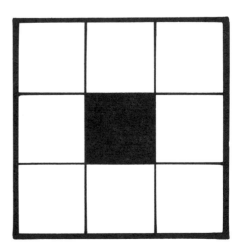

GA1134

Ms. Five

To solve the subtraction problems, count the dots found in each appropriate space. Be careful. There are large and small shapes. List each problem and answer.

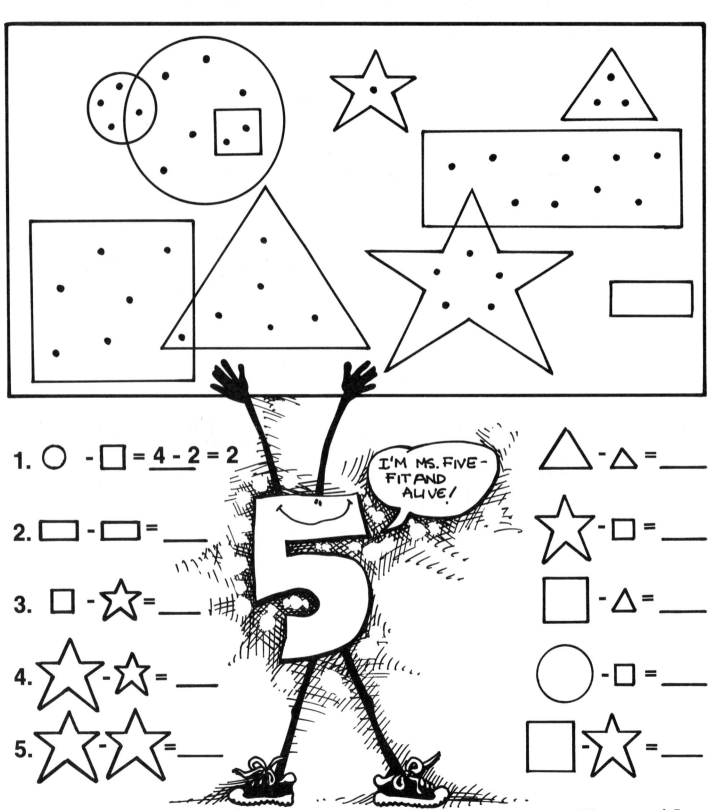

1. ◯ - ▢ = <u>4 - 2</u> = 2

2. ▭ - ▭ = ___

3. ▢ - ☆ = ___

4. ☆ - ☆ = ___

5. ☆ - ☆ = ___

I'M MS. FIVE — FIT AND ALIVE!

△ - △ = ___

☆ - ▢ = ___

▢ - △ = ___

◯ - ▢ = ___

▢ - ☆ = ___

Bonus: Using the shape code, how many subtraction problems with a difference of 5 can you draw?

GA1134

Captain Six

Using the hexagon code, write each subtraction problem and answer. The first one is done for you.

1. $6 - 3 = 3$

2.

3.

4.

5.

6.

7.

Bonus: Can you draw four hexagons using exactly 22 lines?

GA1134

Here's Seven

Complete the chart found below to list and solve each subtraction problem. Example: The first row has a dot under the first digit SEVEN and a dot under the digit ZERO, so the problem is 7 - 0 = 7.

HERE'S SEVEN AND SOME SILLY SIDEKICKS SUBTRACTING.

#	7	6	5	4	3	2	1	0
1. $7 - 0 = 7$	•							•
2.	•	•						
3.							•	•
4.		•					•	
5.	•					•		
6.	•						•	
7.		•				•		
8.	•				•			
9.		•			•			
10.					•	•		
11.						•		•
12.			•	•				
13.			•				•	
14.		•		•				
15.	•			•				
16.				•	•			
17.		•	•					
18.				•			•	
19.			•		•			
20.				•		•		
21.	•		•					
22.					•		•	
23.			•					•
24.			•			•		
25.		•						•
26.					•			•
27.				•				•
28.						•	•	

Bonus: Using the chart above, how many problems can you list with a difference of 1?

GA1134

V-Eight

Time yourself. How many minutes will it take you to race around the track? Begin in the space after START and call out the answers to a friend. Work around the track. Can you complete the race in less than three minutes? Two minutes? Practice until you can do it in less than one minute.

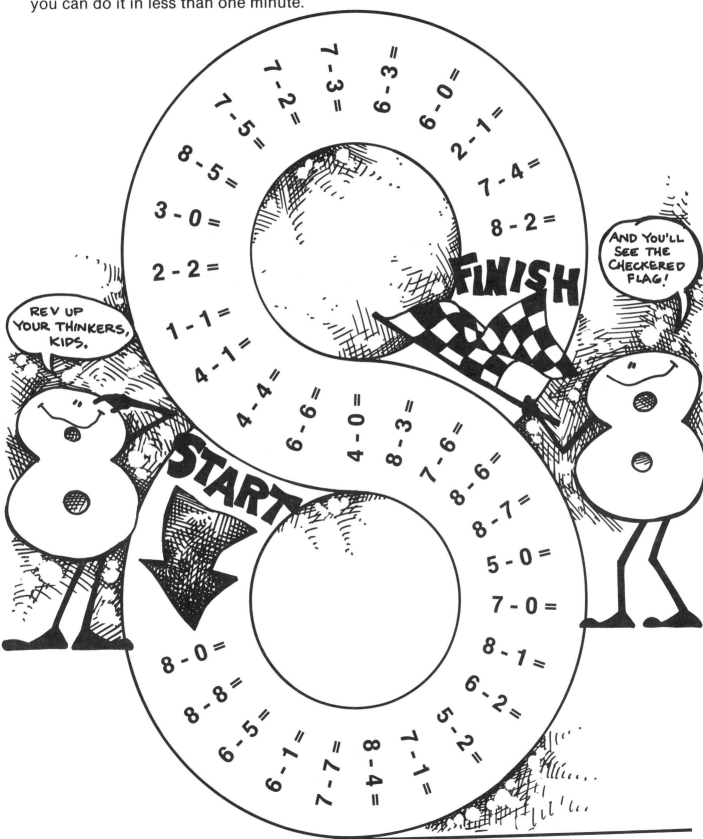

Mr. Nine

Complete the letter code by working some of the problems found below. Then decode the rest of the problems and solve each one. Write the problems and answers and then write the answers in code. Some of the problems have been completed for you.

A = 2   B =   C = 6   D = 4   E =

F = 1

G =

H = 8

I = 9

J =

1. I - I = J
    9 - 9 = 0
2. H - F = B
    8 - 1 = 7
3. B - C = F
4. C - A = E
5. B - F = C
6. I - H = F
7. H - E = E
8. B - D = A
9. H - A =
10. I - F =
11. H - C =
12. I - D =
13. H - B =
14. I - G =

I - A =

C - G =

G - G =

C - E =

I - C =

I - B =

D - D =

H - G =

B - E =

H - D =

B - G =

B - A =

C - C =

C - D =

I FEEL A LITTLE SILLY WITH ALL THESE LETTERS

Bonus: In code, how many different ways can you list problems with a difference of 2?

GA1134

Little Miss Ten

Solve each subtraction problem found in the spaces below. Then color spaces with an even difference YELLOW. Color spaces with an odd difference BLUE.

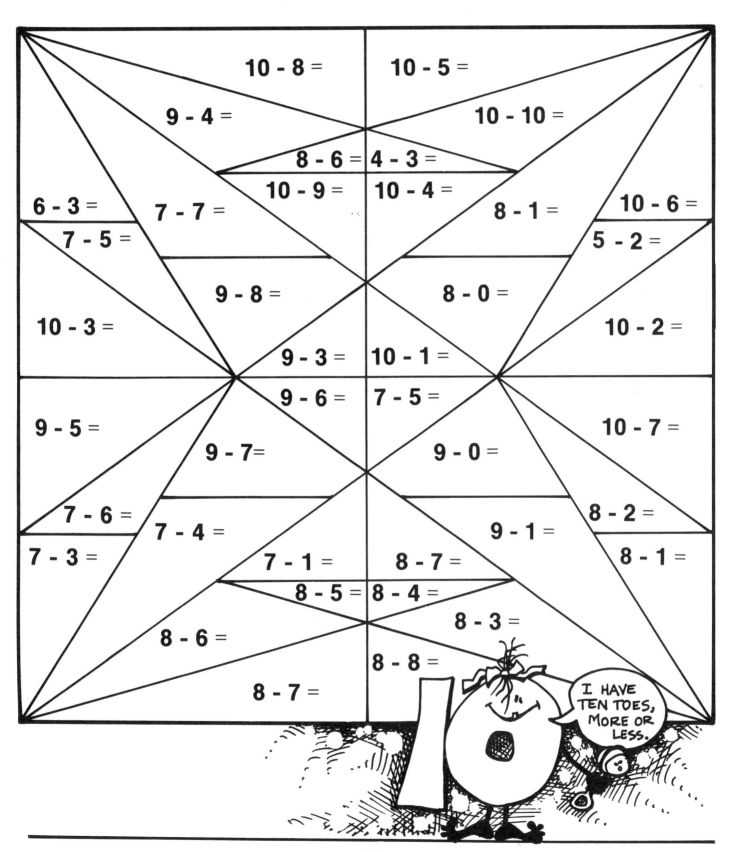

11

GA1134

Subtraction Sunflowers

Complete each sunflower by listing number pairs with the difference shown in the center of each flower. Some of the number pairs have been listed for you.

Bonus: Choose another number and draw a sunflower. Place your number in the center. How many petals will your flower have?

12

GA1134

It's Snowing Number Designs

Write number pairs in the circles connected by a line with a difference of the number shown in the center. One snowflake is done for you.

Bonus: Using only the digits 1-12, how many number pairs can you list with a difference of 2?

13

Number Pairs

Complete each box found below by listing number pairs with a difference indicated by the large number found in the corner of each box. Example: Two number pairs that belong in the first box are 2, 1 and 3, 2 because 2 - 1 = 1 and 3 - 2 = 1.

Bonus: Of all the possible number pairs between 0, 0 and 12, 12, how many have a difference of 0?

Slightly Silly Subtraction Stories

Write the subtraction problem and answer on the line following each story problem found below.

1. Zanny Zak zipped zero marbles inside his jacket pocket while elegant Ellen placed eight marbles in a secret hiding place. How many more marbles does elegant Ellen have than zanny Zak?

   _____

2. Sassy Suzie saved seven slices of spicy salami while sleepy Sam saved only six slices of spicy salami. How many more slices of spicy salami does sassy Suzie have than sleepy Sam?

   _____

3. Flirty Flo found four funny flavored figs in a farmer's hat. Fickle Fran found five funny flavored figs in the farmer's boot. How many more funny flavored figs did fickle Fran find than flirty Flo?

   _____

4. Tiny Tim took ten tangy tangerines to the team's picnic, and terrific Tom took two tangy tangerines to the team's picnic. How many more tangy tangerines did tiny Tim take to the team's picnic than terrific Tom?

   _____

5. Neat Nelly knitted nine nifty nightcaps while her nice, new neighbor Norman, knitted no nifty nightcaps. How many more nifty nightcaps did neat Nelly knit than her nice, new neighbor Norman?

   _____

6. Excited Edward ate every apple in the basket of twelve except those eaten by elfish Esther. If elfish Esther ate eleven apples, how many did excited Edward eat?

   _____

Bonus: Tired Thomas played tic-tac-toe with tickled Tina. Tired Thomas won all but three games against tickled Tina. If tired Thomas and tickled Tina played twelve terrific tic-tac-toe games, how many games did tickled Tina win?

GA1134

"X" Marks the Code

Use the code to complete the problems. Example: $\wedge$ = 8 and $\dot{\wedge}$ = 12

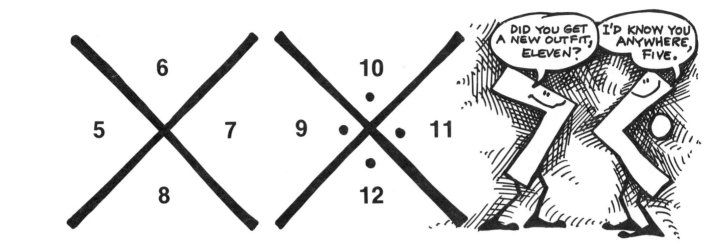

1. $\wedge - \,\rangle\; =$ $\vee - \,\rangle\; =$ $\langle - \langle\; =$

2. $\langle\!\cdot - \vee\; =$ $\dot\rangle - \langle\; =$ $\dot\rangle - \vee\; =$

3. $\dot\rangle - \wedge\; =$ $\dot\vee - \,\rangle\; =$ $\dot\wedge - \dot\vee\; =$

4. $\dot\wedge - \wedge\; =$ $\langle\!\cdot - \langle\; =$ $\langle - \vee\; =$

5. $\langle\!\cdot - \,\rangle\; =$ $\wedge - \vee\; =$ $\dot\rangle - \dot\rangle\; =$

6. $\dot\wedge - \dot\vee\; =$ $\dot\wedge - \vee\; =$ $\vee - \vee\; =$

7. $\langle - \,\rangle\; =$ $\dot\rangle - \,\rangle\; =$ $\dot\wedge - \langle\; =$

8. $\rangle - \,\rangle\; =$ $\dot\wedge - \,\rangle\; =$ $\dot\wedge - \dot\rangle\; =$

Bonus: Write a subtraction problem with the largest possible difference using the "X" code.

   GA1134

Sea Serpent Subtraction

Complete the subtraction problem found in each of the sea creature's spots. Color the spots with a difference of two YELLOW. Color the spots with a difference of three BLUE. Color the spots with a difference of four PURPLE. Color the spots with a difference of five ORANGE.

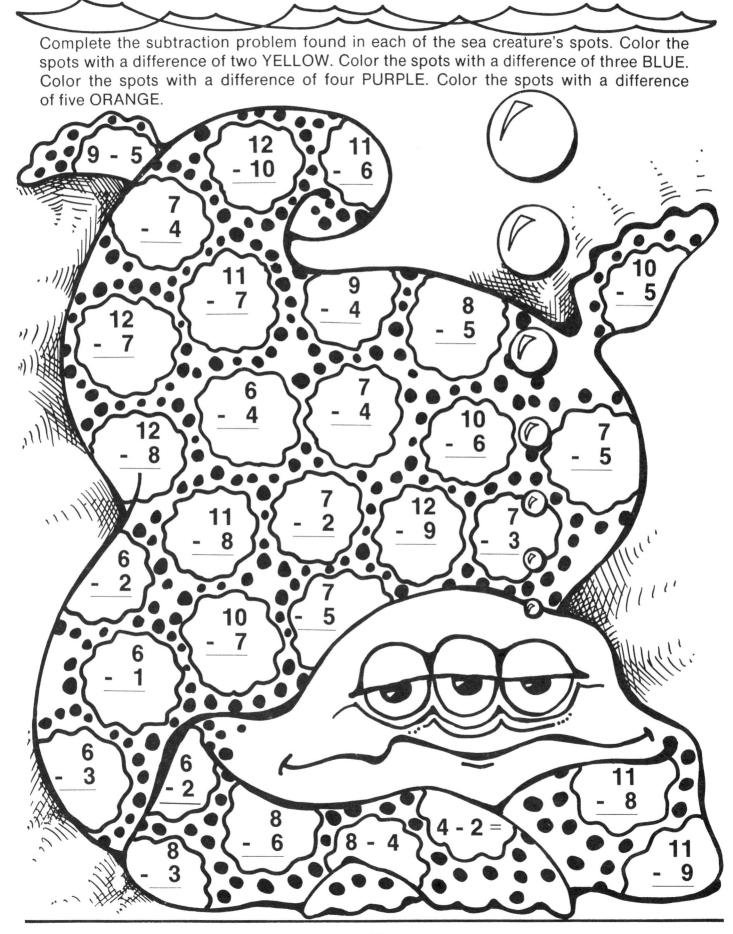

17

Signing Subtraction

Use the sign language alphabet to solve the subtraction problems found below.

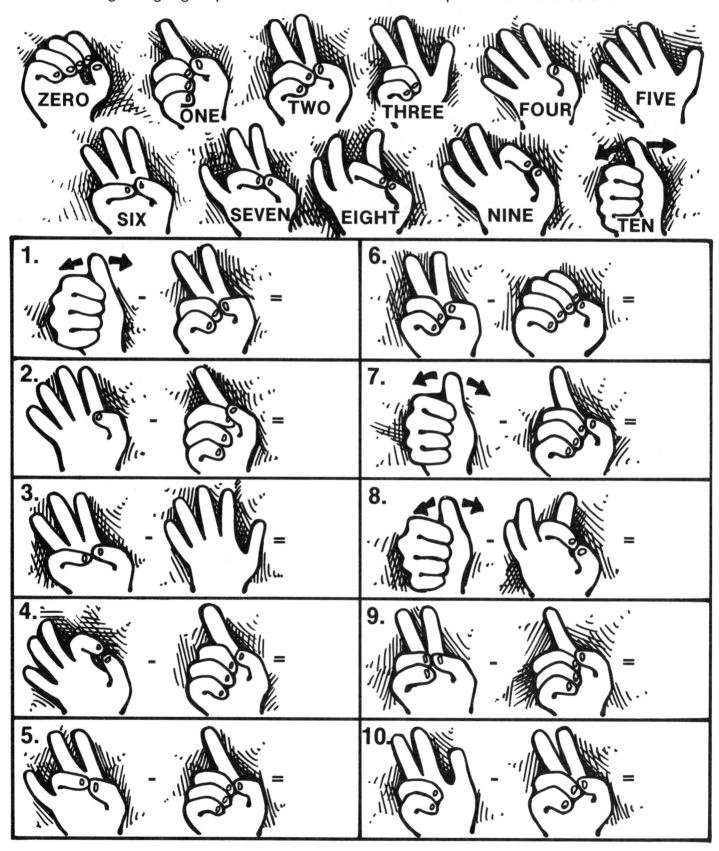

Bonus: Learn to sign your age, phone number or address.

GA1134

Art with a Difference

Complete each subtraction problem found in the design below.
Color sections of the design with a difference of 1 PINK.
Color sections of the design with a difference of 2 YELLOW.
Color sections of the design with a difference of 3 RED.
Color sections of the design with a difference of 4 ORANGE.

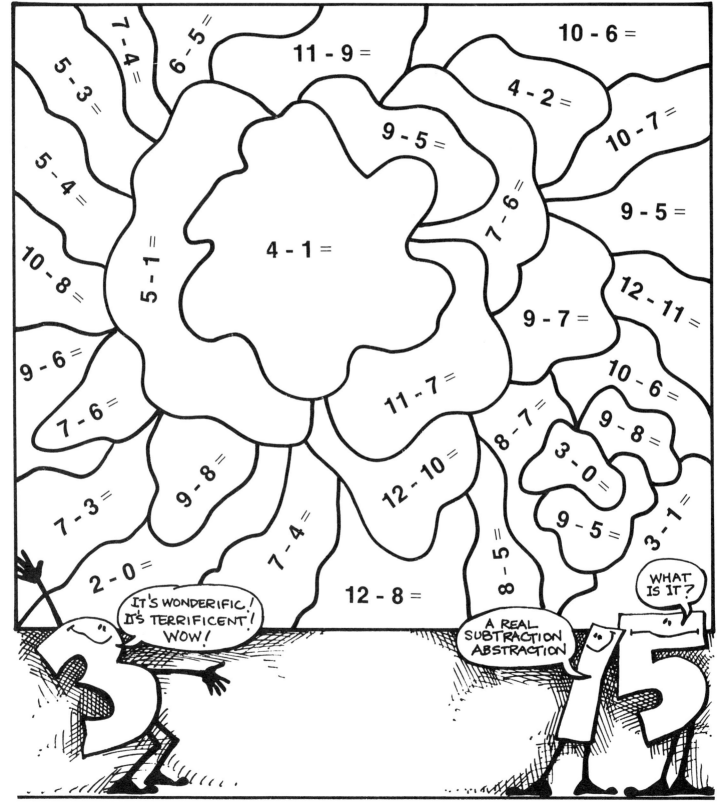

GA1134

Subtraction Bull's-Eye

What number goes in each center bull's-eye and in the other empty spaces? The outside ring is the answer when you subtract the inside ring from the bull's-eye.

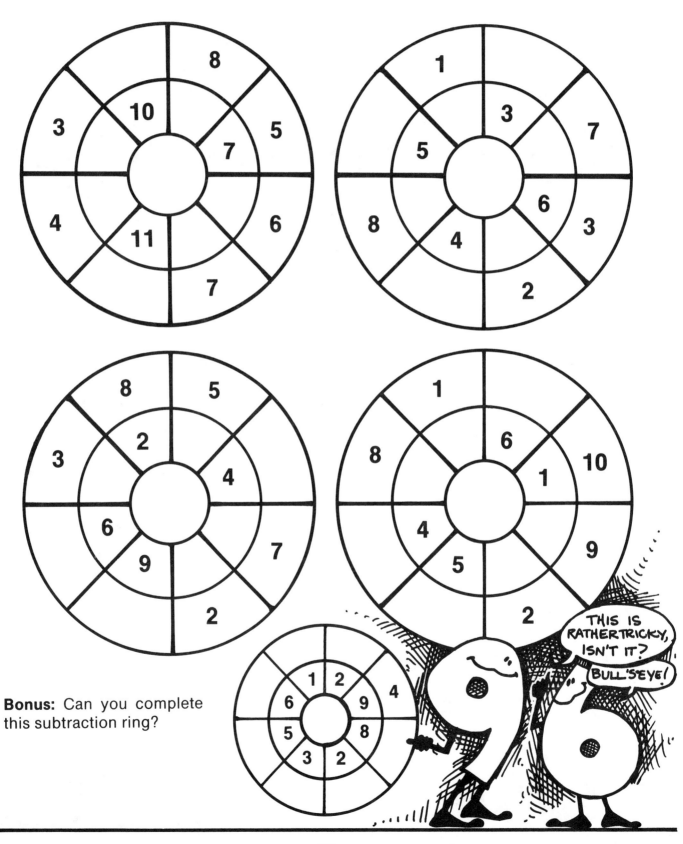

Bonus: Can you complete this subtraction ring?

20

GA1134

Down the Path of Subtraction

Begin with the number at the top, and follow the path subtracting each number you cross. Write the answer in the circle at the end of each path.

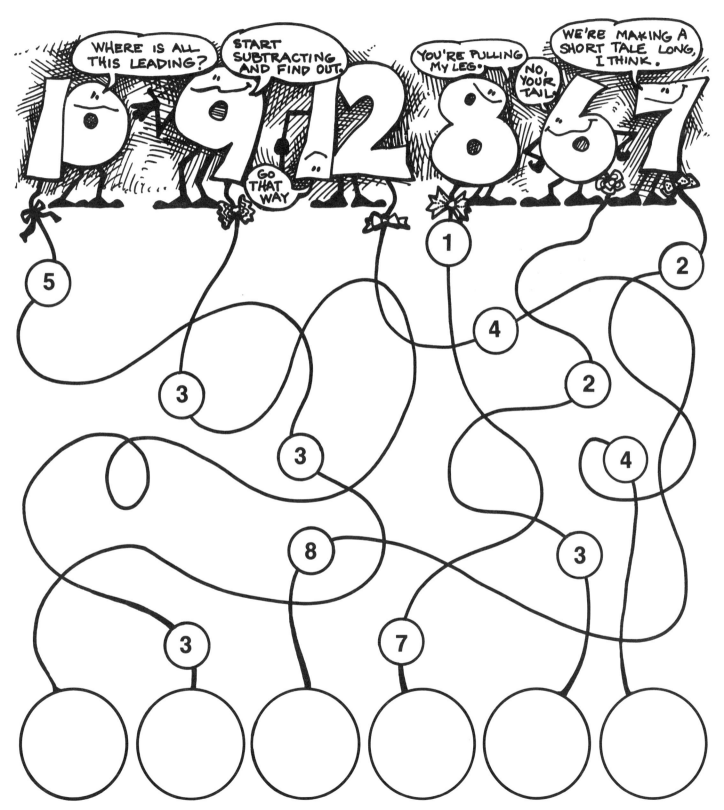

Bonus: Create your own subtraction path puzzle for a friend.

Domino Differences

Use the complete set of dominoes found at the bottom of this page to answer each question. Always subtract the smaller number from the larger number.

Example: [domino] and [domino] are the same problem.   6 - 1 = 5

1. How many dominoes have a difference of 0?
2. How many dominoes have a difference of 1?
3. How many dominoes have a difference of 2?
4. How many dominoes have a difference of 3?
5. How many dominoes have a difference of 4?
6. How many dominoes have a difference of 5?
7. How many dominoes have a difference of 6?

0 - 0 = 0   1 - 0 = 1   1 - 0 = 1   2 - 0 = 2

Bonus: Which difference is the most common? Which difference is the least common?

Checking the Facts

For each row of checks found below, list and solve the subtraction problem. The first one has been completed for you.

#		12	11	10	9	8	7	6	5	4	3	2	1
1.	12 – 10 = 2	✔		✔									
2.			✔	✔									
3.			✔						✔				
4.			✔							✔			
5.		✔							✔				
6.			✔										✔
7.		✔				✔							
8.		✔					✔						
9.				✔									✔
10.				✔					✔				
11.			✔					✔					
12.		✔					✔						
13.		✔											✔
14.				✔			✔						
15.				✔		✔							
16.				✔							✔		
17.				✔				✔					
18.		✔				✔							
19.				✔	✔								
20.			✔					✔					
21.					✔	✔							
22.		✔			✔								
23.					✔		✔						

Bonus: What is the answer when you subtract all the little numbers from their mamas?

23

GA1134

Circle Three

Find and circle exactly ten combinations of three digits that are true subtraction sentences in each box found below. Every number will be circled once and only once. Some of the subtraction sentences have been circled for you.

1.

```
 6   3   3    5   2   3
 4   9   6    7   1   6
 2   5   5    8   1   7
 2   4   1    8   4   4
 9   6   3   10   3   7
```

3.

```
10  10   7    7   0  11
 1   6   8    3   5   4
 9   4   7    6   1   7
 6   0   6   11   9   2
 6   6   0   12   0  12
```

2.

```
 7   3   4   10   5   5
 6  10   3    7  10  11
 2  11   7    4   9   3
 4  12   3    9   1   8
 5   3   2    9   8   1
```

4.

```
12  11   4    3   1   6
 6   5   5    3   2   1
 6   6  11   10   1   5
11   6   5   11  11   0
11   8   3   11   1  10
```

Bonus: Create your own circle-three puzzle.

GA1134

Subtraction Tower

Start at the bottom of the tower and subtract your way up. Each row is built on the answers you get when you subtract the two numbers under it. Some numbers have been filled in for you.

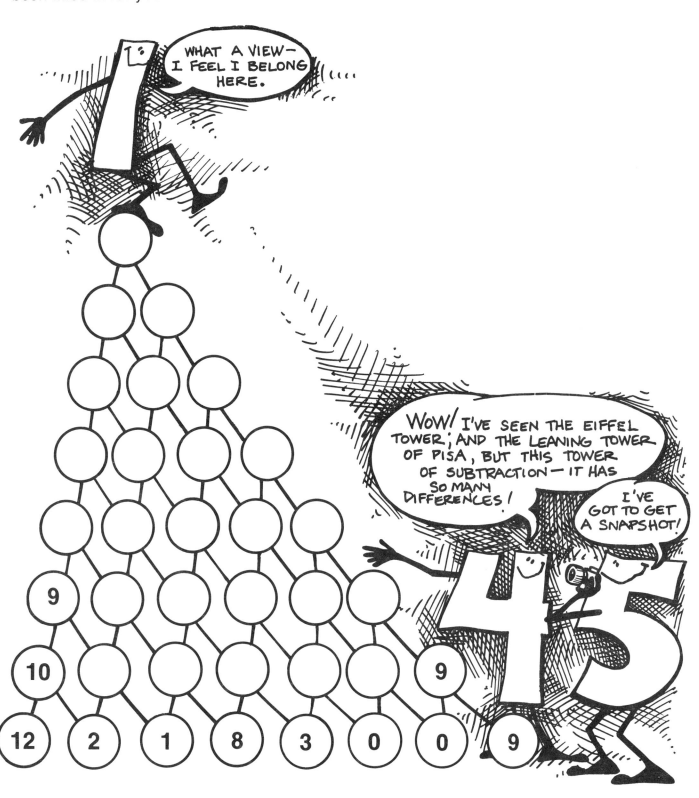

Bonus: Create your own subtraction tower with seven circles at the bottom. Place the digits of your phone number in the circles on the bottom of your tower.

GA1134

Subtraction Quilt

Subtract to find the difference between each row and column and write the answer in each quilt square. For example, the number in the first square is 0 because 1 – 1 = 0.

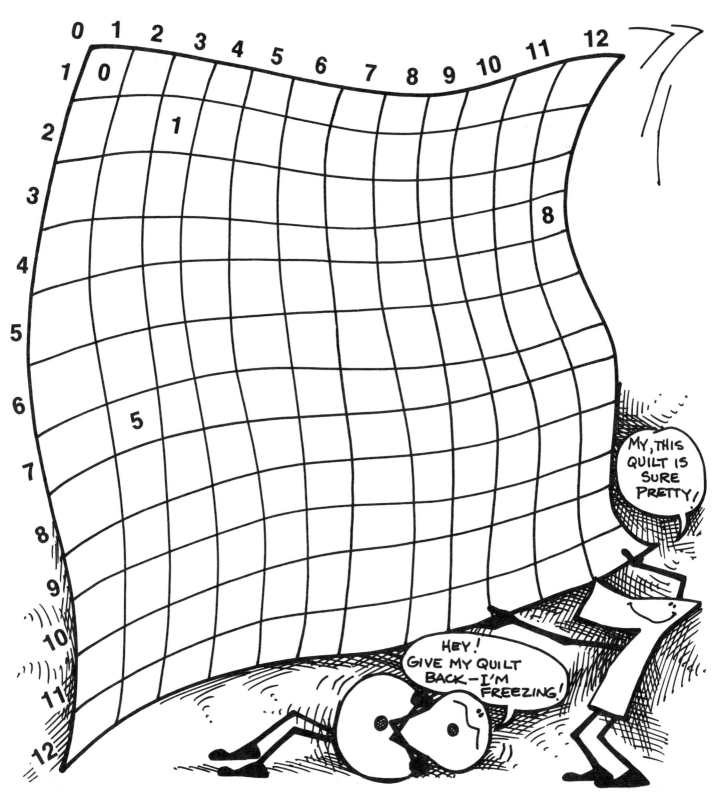

Bonus: Color all the spaces containing a zero the same color. Color all the spaces containing a one the same color.

GA1134

To discover the Secret Message, work each subtraction problem found below. Then fill in each blank with the appropriate letter. Example: 9 - 8 = 1, so place the letter A in each blank marked with the number 1.

A. 9 - 8 =

B. 12 - 0 =

C. 7 - 5 =

D. 15 - 2 =

E. 7 - 4 =

F. 15 - 1 =

G. 6 - 2 =

H. 20 - 4 =

I. 10 - 5 =

J. 12 - 1 =

K. 12 - 6 =

L. 10 - 1 =

M. 11 - 4 =

N. 15 - 0 =

P. 20 - 3 =

R. 11 - 1 =

T. 20 - 1 =

O. 11 - 3 =

S. 20 - 2 =

Y. 25 - 4 =

U. 25 - 5 =

Secret Message:

$\overline{\;\;}_{5}\;\overline{\;\;}_{14}$ $\overline{\;\;}_{21}\;\overline{\;\;}_{8}\;\overline{\;\;}_{20}$ $\overline{\;\;}_{9}\;\overline{\;\;}_{3}\;\overline{\;\;}_{1}\;\overline{\;\;}_{10}\;\overline{\;\;}_{15}$ $\overline{\;\;}_{19}\;\overline{\;\;}_{16}\;\overline{\;\;}_{3}$ $\overline{\;\;}_{7}\;\overline{\;\;}_{1}\;\overline{\;\;}_{19}\;\overline{\;\;}_{16}$

$\overline{\;\;}_{14}\;\overline{\;\;}_{1}\;\overline{\;\;}_{2}\;\overline{\;\;}_{19}\;\overline{\;\;}_{18}$, $\overline{\;\;}_{18}\;\overline{\;\;}_{20}\;\overline{\;\;}_{12}\;\overline{\;\;}_{19}\;\overline{\;\;}_{10}\;\overline{\;\;}_{1}\;\overline{\;\;}_{2}\;\overline{\;\;}_{19}\;\overline{\;\;}_{5}\;\overline{\;\;}_{8}\;\overline{\;\;}_{15}$

$\overline{\;\;}_{2}\;\overline{\;\;}_{1}\;\overline{\;\;}_{15}$ $\overline{\;\;}_{12}\;\overline{\;\;}_{3}$ $\overline{\;\;}_{14}\;\overline{\;\;}_{20}\;\overline{\;\;}_{15}$.

Bonus: Write a subtraction problem for each number of your telephone number.

GA1134

Be Mine Subtraction

Study the hearts carefully; then complete each problem below by placing each number by its heart. The first one is done for you.

1. =
   8 - 2

   - =
   7 - 1

   - =
   7 - 4

2. - =
   - =
   - =

3. - =
   - =
   - =

4. - =
   - =
   - =

5. - =
   - =
   - =

Bonus: Write your phone number using this code.

28

Coded Subtraction

Use the code to write and solve each subtraction problem found below.

(speech bubbles: "THESE AREN'T NUMBERS — THEY'RE SHAPES!" "DON'T YOU KNOW A CODE WHEN YOU SEE ONE?" "I FOR '1' THINK THEY'RE CUTE.")

0 = ○		4 = ⊛		8 = ⊠	
1 = ⊘		5 = □		9 = ⊞	
2 = ⊗		6 = ◺		10 = △	
3 = ⊗		7 = ⊠		11 = △	

Column 1

1. △ − △ =
2. △ − ⊗ =
3. ⊠ − ◺ =
4. ◺ − ◺ =
5. △ − ⊛ =
6. △ − ⊠ =
7. △ − ⊗ =
8. △ − ⊠ =
9. △ − □ =
10. ⊠ − ⊠ =
11. ⊞ − ⊠ =
12. △ − ⊛ =

Column 2

1. ⊠ − ○ =
2. ⊠ − □ =
3. ⊠ − ⊗ =
4. △ − ⊞ =
5. ⊠ − ⊗ =
6. △ − ◺ =
7. ⊠ − ⊠ =
8. △ − □ =
9. △ − ⊛ =
10. △ − ⊠ =
11. □ − □ =
12. △ − ⊗ =

Column 3

1. ⊠ − ⊘ =
2. △ − ○ =
3. ⊠ − ⊛ =
4. △ − ◺ =
5. ⊠ − ⊗ =
6. ⊞ − ⊗ =
7. △ − ⊠ =
8. ⊞ − ⊠ =
9. ⊠ − ⊗ =
10. △ − ⊠ =
11. ⊠ − ⊘ =
12. ⊠ − ○ =

Bonus: Look for a pattern in the code and extend it so you can solve the following problem: △ − △ =

GA11

Tic-Tac Subtract

Example:

$\Box - \Box = 8 - 7 = 1$

Use the code to solve each addition problem.

1. $\Box - \Box = 11 - 10 = 1$ $\Box - \Box =$ $\Box - \Box =$

2. $\Box - \Box =$ $\Box - \Box =$ $\Box - \Box =$

3. $\Box - \Box =$ $\Box - \Box =$ $\Box - \Box =$

4. $\Box - \Box =$ $\Box - \Box =$ $\Box - \Box =$

5. $\Box - \Box =$ $\Box - \Box =$ $\Box - \Box =$

6. $\Box - \Box =$ $\Box - \Box =$ $\Box - \Box =$

7. $\Box - \Box =$ $\Box - \Box =$ $\Box - \Box =$

8. $\Box - \Box =$ $\Box - \Box =$ $\Box - \Box =$

9. $\Box - \Box =$ $\Box - \Box =$ $\Box - \Box =$

10. $\Box - \Box =$ $\Box - \Box =$ $\Box - \Box =$

Bonus: How many problems can you write in code that have a difference of 4?

GA1134

Subtraction Line Design #1

Use a ruler to draw a line connecting each number pair that has a difference of 4.

1

8   2

7   3

6   4

5   5

4   6

3   7

2   8

Bonus: Color your line design.

GA1134

Subtraction Line Design, Too

Use a ruler to draw a line connecting each number pair that has a difference of 4, 5 or 10.

•13

5•

.12

6.

•11

7.

10

8

9

Bonus: Draw a circle. Place numbers around the edges and create your own subtraction line design.

GA1134

Search and Circle

Find and circle any true subtraction sentences. Example: 10, 8 and 2 are ringed in the top left-hand row because 10 - 8 = 2. Can you find and circle 30 subtraction sentences? 40? 50?

(10	8	2)	3	11	9	2	11	7	9	11	2	4
11	8	3	7	5	2	8	10	4	8	6	1	4
7	1	7	10	7	3	2	1	3	1	5	3	8
4	7	2	1	6	12	3	9	9	3	6	1	0
12	0	12	9	4	10	8	6	0	5	2	7	4
10	2	6	6	0	2	8	5	9	5	4	8	6
9	3	10	6	4	8	0	10	3	7	4	6	2
1	4	5	6	11	12	8	4	11	11	0	1	9
11	6	5	3	3	9	6	3	9	12	6	5	11
1	12	7	5	8	2	8	4	4	8	5	3	10
10	9	6	11	4	7	6	9	5	4	1	2	1
12	1	4	5	0	11	4	7	8	6	2	3	3
9	8	2	6	12	6	6	2	5	3	4	7	3
3	12	4	8	12	5	7	7	3	8	2	10	1

WE'RE A CIRCLE OF FRIENDS.

RING AROUND THE MATH FACTS!

Bonus: Make up your own hidden subtraction puzzle. Give it to a friend to solve.

33

GA1134

Rolling Differences

Roll a pair of dice one hundred times. Each time record the difference between the number of dots on each die. Example: If you roll a one and a six, the difference is five, and you would put a tally mark beside the number five. Tally marks look like this: 5 = |||| 12 = |||| |||| ||

Differences

0	
1	
2	
3	
4	
5	

YOU ROLL THE DICE... I'LL TALLY

WE'RE GETTING DIZZY!!!

What difference did you roll the most times?

What difference did you roll the least times?

Bonus: Which two differences should be rolled approximately the same number of times out of 36? If you roll the dice 36 times, approximately half of the time one of two numbers will be the difference. What are those two numbers?

Sub Matchup

To discover a good way to master math facts, draw a line connecting each number pair in the left-hand column with a number pair in the right-hand column that has the same difference. Then write the letters that are intersected by a line in the order they are found from top to bottom.

Mystery Word:

12 - 6 = •

10 - 1 = •

11 - 1 = •

9 - 2 = •

11 - 10 = •

12 - 4 = •

12 - 1 = •

12 - 7 = •

8 - 4 = •

12 - 9 = •

12 - 12 = •

12 - 10 = •

(M) (P) (E) (R) (A) (M) (O) (C) (T) (R) (I) (I) (Z) (E) (C) (E)

• 9 - 0 =

• 11 - 4 =

• 10 - 4 =

• 9 - 1 =

• 12 - 2 =

• 12 - 11 =

• 11 - 7 =

• 7 - 2 =

• 11 - 0 =

• 9 - 7 = 2

• 10 - 7 =

• 10 - 10 =

Bonus: Write down the unused letters to spell yet another way to master math facts.

GA1134

Computer Subtraction

Using the number found at the beginning of each row below as input, complete the subtraction problem using the flow chart. The first one has been completed for you.

1.	5	(five is odd)	5 - 2 = 3 + 3 = 6
2.	3		
3.	4		
4.	12		
5.	10		
6.	7		
7.	11		
8.	9		
9.	8		
10.	6		
11.	13		
12.	2		

Bonus: Change EVEN to ODD and ODD to EVEN on the flow chart and rework each problem.

Arrow Subtraction

Use the chart found below to complete the subtraction problems. The arrow tells you what direction to look on the chart to find the number to be subtracted from each given number. Example: 9➜ = 9 - 4. The arrow stands for the number on the right side of the nine.

1. ↖9

2. ↑8

3. ↑9

4. 9↓

5. 9↗

6. ↙9

7. 8↓

8. 9↘

9. 8↗

10. 8↘

11. ↑5

12. ↑4

13. 7➜

14. ←3

15. ←2

Bonus: Rearrange the digits 1-9 in the spaces and do ten subtraction problems using the new arrow code.

GA1134

Same Difference

Hidden in each row of digits below are number pairs with a difference of the first number in that row. Find and circle the number pairs. The first row has been done for you.

1.	1	5 5 (8 7 6) 3 9 4 (9 8 7)
2.	3	5 2 3 9 6 3 4 6 3 8 5
3.	5	9 5 0 9 4 8 3 6 2 7 2
4.	2	9 8 6 5 3 1 2 0 7 5 3
5.	6	7 1 8 3 9 3 6 0 7 1 8
6.	4	6 2 8 4 7 3 5 1 6 2 4
7.	8	9 1 7 9 8 0 9 8 2 8 0
8.	9	9 0 8 1 5 4 3 6 5 2 7
9.	7	9 2 6 4 7 0 8 1 5 2 4
10.	0	5 5 6 7 8 8 4 4 3 3 7

WE COUPLES ARE ALL RELATED.

YES, WE ALL HAVE THE SAME DIFFERENCES.

Bonus: Make up your own hidden number pair puzzles for the digits 10, 11 and 12.

38

Follow the Leader

Complete the chart found below by following the directions at the top of each column. The first row has been completed for you.

Begin	SUBTRACT FROM ME.	ADD ME.	SUBTRACT ME.	ADD ME AND PUT THE ANSWER HERE.	Answer
1. 7	12 - 7 = 5	5 + 5 = 10	10 - 2 = 8	8 + 1 =	9
2. 5					
3. 2					
4. 6					
5. 10					
6. 4					
7. 3					
8. 12					
9. 8					
10. 9					
11. 11					
12. 1					

Bonus: Change the directions at the top of the first column to read: SUBTRACT FROM 20 and solve each problem again.

GA1134

Plenty of Twenty

1. Pick any number in the twenties and write it in the first box. (25)
2. Add its two digits. (2 + 5 = 7)
3. Subtract the sum of the two digits from the original number you chose. (25 - 7 = 18)
4. Repeat steps 1-3 five more times, once in each box, choosing a different number each time.

Bonus: What pattern did you discover?

GA1134

Subtraction-Go-Round

Arrange the digits 1-9 in the circles so that each of the four lines of three circles is a true subtraction sentence.

Bonus: Arrange the even digits 2-18 in the circles so that each of the four lines of three circles is a true subtraction sentence.

Circle Search

Hidden in each circle are enough numbers to make a true subtraction sentence. Can you use all the digits in each circle to create a subtraction sentence? The first one has been completed for you.

1. 6 - 3 = 3 _____   _____   _____   _____

2. _____   _____   _____   _____

3. _____   _____   _____   _____

4. _____   _____   _____   _____

Bonus: Can you write three different true subtraction sentences using all five of these digits: 0, 1, 1, 2, 2?

GA1134

Down the Path

Begin at the top number and trace the path, subtracting the two numbers found in circles along each path. Write each answer in the empty circle found at the end of each path.

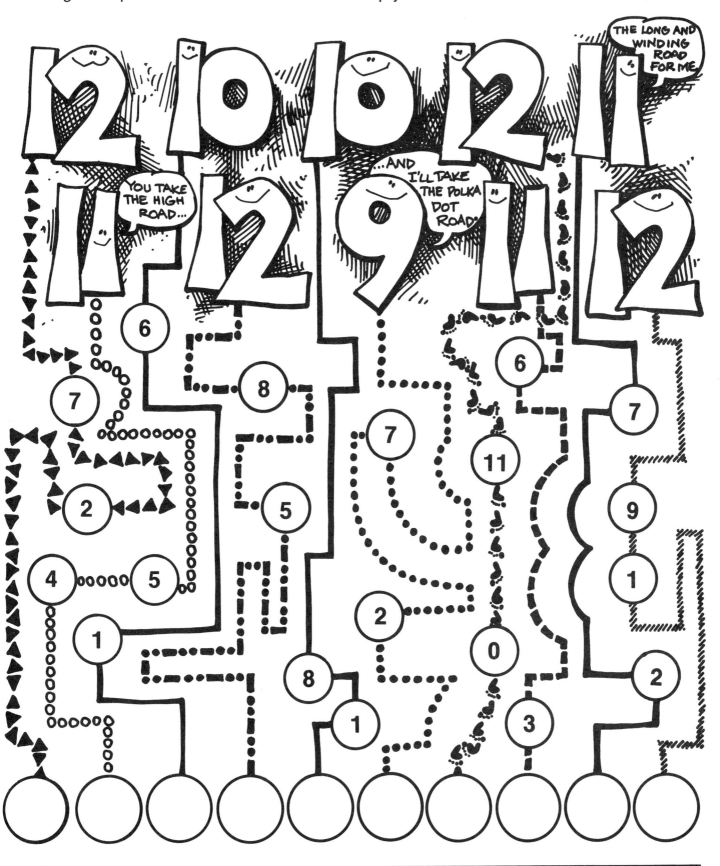

Cut the Cake

Draw two straight lines to divide the cake so that each area has exactly two numbers with a difference of 2.

Draw two straight lines to divide the cake so that each area has exactly two numbers with a difference of 3.

Draw two straight lines to divide the cake so that each area has exactly two numbers with a difference of 4.

Draw two straight lines to divide the cake so that each area has exactly two numbers with a difference of 5.

GA1134

Shamrock Subtraction

To find out why Ireland will someday be the most populated country in the world, solve each subtraction problem found below. Then place the appropriate letters in the numbered blanks.

T 28 - 10 = 18	R 16 - 8	Q 26 - 1	O 11 - 5	N 12 - 2
A 9 - 6	S 14 - 7	P 24 - 0	M 12 - 3	L 24 - 12
U 8 - 6	B 33 - 10	I 12 - 7	J 24 - 11	K 22 - 11
G 28 - 13	H 15 - 1	C 33 - 11	D 32 - 11	X 29 - 10
F 29 - 12	E 10 - 6	V 27 - 10	Y 20 - 0	W 8 - 7

WITH THE "LUCK O' THE IRISH" YOU CAN SOLVE THIS SECRET MESSAGE.

LUCK AND SUBTRACTION.

Secret Message:

☐ ☐ ☐ ☐ ☐ ☐ ☐
23 4 22 3 2 7 4

☐ ☐ ☐   ☐ ☐ ☐ ☐ ☐ ☐ ☐
5 18 7   22 3 24 5 18 3 12

☐ ☐ ☐   ☐ ☐ ☐ ☐ ☐ ☐   ☐ ☐ ☐ ☐ ☐ ☐
5 7   3 12 1 3 20 7   21 2 23 12 5 10

Bonus: Write new subtraction problems with differences that spell your first name.
Example: JIM = 13 - 0 = J, 10 - 5 = I, 11 - 2 = M

GA1134

Hide-and-Seek Nines

1. Choose any number with three different digits. (345)
2. Reverse the numbers. (345 becomes 543)
3. Subtract the smaller number from the larger number. (543 - 345 = 198)
4. Circle the center digit of the answer. (1⑨8)
5. Add the first and last digits. (1 + 8 = 9)
6. Repeat steps 1-5 five more times, once in each box using a different number with three different digits each time.

Bonus: What pattern did you discover?

GA1134

Nine Is Divine

1. Choose any two-digit number. (82)
2. Add the digits of the number. (8 + 2 = 10)
3. Subtract the sum from the original number you chose. (82 - 10 = 72)
4. Add the digits of the answer. (7 + 2 = 9)
5. Repeat steps 1-4 five more times, once in each box, choosing a different two-digit number each time.

Bonus: What pattern did you discover?

GA1134

Nutty Nifty Number Nine

1. Pick any number with four different digits. (3451)
2. Scramble the four digits any way you like. (4513)
3. Subtract the smaller number from the larger number. (4513 - 3451 = 1062)
4. Add all the digits in the answer. (1 + 0 + 6 + 2 = 9) If the answer is a two-digit number, add those digits together.
5. Repeat steps 1-4 five more times, once in each box, using different four-digit numbers.

Bonus: What pattern did you discover?

Pick Four Numbers

1. Choose any four-digit number that has the first and last digit that differs by more than one. Example: 3245 is OK because 5 - 3 is more than 1. 3242 is not OK because 3 - 2 is 1.
2. Reverse the first and last digits and write the new number. (3245 becomes 5243)
3. Subtract the smaller number from the larger number. (5243 - 3245 = 1998)
4. Reverse the first and last digits of the answer. (1998 becomes 8991)
5. Add these two numbers. (1998 + 8991 = 10,989)
6. Repeat steps 1-5 five more times, once in each box, choosing a different four-digit number each time.

Bonus: What pattern did you discover?

GA1134

Subtraction Patterns

Find the subtraction pattern in each row of numbers found below. Continue each row of numbers to include three more numbers. Then write the number pattern on the line beside each. Example: The pattern for the first one is subtract 1, subtract 2, subtract 1, subtract 2, etc.

1. | 12 | 11 | 9 | 8 | 6 | | | |

2. | 20 | 19 | 16 | 15 | 12 | | | |

3. | 12 | 10 | 10 | 8 | 8 | | | |

4. | 20 | 18 | 15 | 13 | 10 | | | |

5. | 24 | 18 | 13 | 9 | 6 | | | |

6. | 25 | 20 | 19 | 14 | 13 | | | |

7. | 50 | 40 | 38 | 28 | 26 | | | |

8. | 30 | 25 | 21 | 16 | 12 | | | |

Bonus: Can you add three more numbers to this number pattern?

| 100 | 99 | 97 | 94 | 90 | | | |

GA1134

Home Run Subtraction

ANNE 16	JAMES 15	DONNA 10
BRYAN 13	VALERIE 11	VICTOR 9

1. Which two children have a difference of seven home runs?

   _____

2. Which is greater, the difference between Anne's and Valerie's home runs or the difference between Donna's and James' score?

   _____

3. What is the difference between the girls' total home runs and the boys' total home runs?

   _____

4. What is the difference between Bryan's and Anne's scores?

   _____

5. Which two boys have a difference of six home runs?

   _____

6. Which two girls have a difference of only one home run?

   _____

Bonus: If one of the girls had scored half as many home runs and one of the boys had scored twice as many home runs, the difference between the girl's total home runs and the boy's total home runs would be twenty points. Which boy's and which girl's scores would have to be changed to make this point difference?

GA1134

Give Me a Sign

If you put the subtraction sign and the equal sign in the right place in each group of digits found below, you will make a true subtraction sentence. Example: 5 5 0 is 5 - 5 = 0 and 7 9 2 is 7 = 9 - 2.

1.	4	1	3			1	2	1	1	1
2.	8	1	1	3		1	2	1	2	0
3.	3	1	1	8		1	2	1	0	2
4.	4	3	1			4	2	2		
5.	5	2	3			1	1	7	4	
6.	1	5	4			4	1	2	8	
7.	5	6	1			1	1	8	3	
8.	1	2	0	1	2	9	1	1	2	
9.	1	1	1	2	1	0	4	4		
10.	1	1	1	0	1	0	1	0	1	0
11.	2	1	1	9		3	1	2	9	
12.	1	2	2	1	0	1	1	6	5	
13.	9	1	2	3		1	1	1	10	
14.	1	2	6	6		8	1	2	4	
15.	5	1	2	7		7	1	2	5	
16.	1	1	5	6		1	1	1	1	0

Bonus: Create six of your own give-me-a-sign puzzles.

GA1134

Two Signs, Please

Place two signs, a subtraction sign and an equal sign, in each row of digits to make true subtraction sentences. The first one has been completed for you.

1. 6 9 – 5 = 6 4

2. 7 2 1 2 6 0

3. 3 8 2 0 1 8

4. 4 8 7 4 1

5. 9 7 1 4 8 3

6. 6 9 9 6 0

7. 7 4 2 3 5 1

1 1 4 7 1 0 7

1 1 0 1 0 1 0 0

1 8 9 4 5 1 4 4

1 7 9 1 7 2 7

1 8 7 1 0 7 8 0

6 0 9 6 0 0 9

1 4 7 4 7 1 0 0

Bonus: 1 0 0 9 – 9 1 0 0 0

GA1134

Connecting Pairs

Draw lines connecting number pairs with a difference of 6. You must not cross any line. When you finish, every dot should be connected to a number pair. One path has been completed for you.

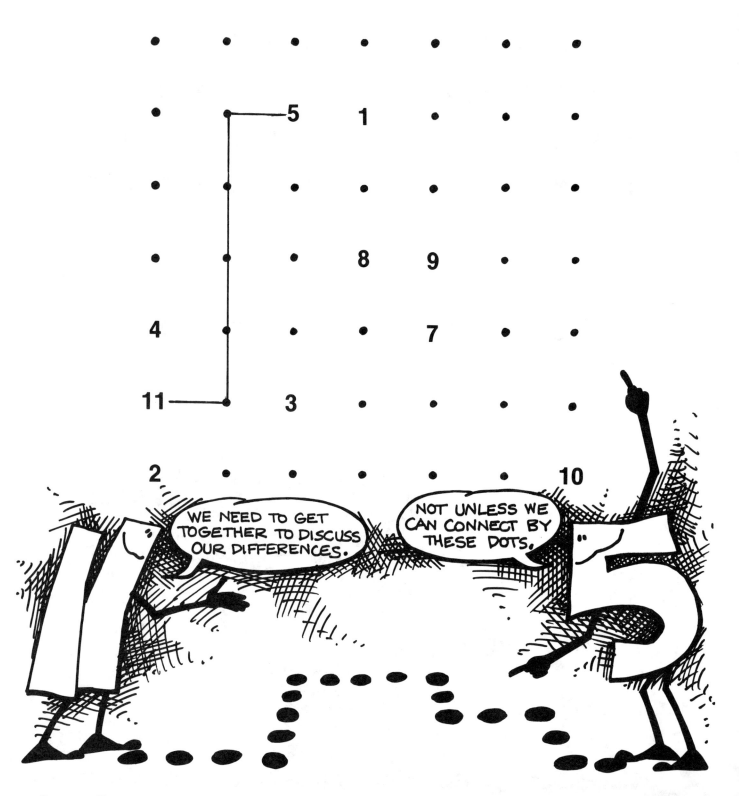

Bonus: Draw your own puzzle with number pairs that differ by 5. Give your puzzle to a friend to solve.

1. If you pay for a peppermint with a dime, how much change will you get?

2. If you pay for a bag of gumdrops with a quarter, how much change will you get?

3. If you pay for a candy stick with two dimes, how much change will you get?

4. If you pay for a peppermint with a quarter, how much change will you get?

5. If you pay for a bag of gumdrops with a dime, how much change will you get?

6. If you pay for a chocolate bar with two dimes, how much change will you get?

7. If you pay for a candy stick with a quarter, how much change will you get?

8. If you pay for a peppermint and a bag of gumdrops with a quarter, how much change will you get?

9. If you pay for a candy stick and a chocolate bar with two quarters, how much change will you get?

10. If you buy one of each and pay for it with fifty cents, how much change will you get?

Bonus: If you have a dollar and you and your friend want one of each kind of candy, do you have enough to pay for it?

GA1134

Subtraction Check

The answer for each problem below is given. You must list the number pair with the appropriate difference for each. Some of the numbers have been checked for you, others you will have to figure out for yourself. Study the example in the first row and then complete the chart, listing the equation and placing the proper checks in each row.

		12	9	8	7	3	1
1.	6 = 7 - 1				✓		✓
2.	11 =						
3.	9 =						
4.	8 =						
5.	7 =						
6.	4 =				✓		
7.	6 =		✓				
8.	5 =	✓					
9.	5 =					✓	
10.	2 =		✓				
11.	2 =						✓
12.	1 =		✓				

Bonus: Draw a chart just like this one. At the top of your chart, list the numbers 11, 10, 6, 5, 4, 2. How many check subtraction problems can you make?

Making Special Arrangements

1. Arrange three 4's to make a subtraction problem with a difference of 40.

2. Arrange five 6's to make a subtraction problem with a difference of 600.

2. Arrange four 7's to make a subtraction problem with a difference of 770.

4. Arrange seven 9's to make a subtraction problem with a difference of 9000.

Bonus: Can you arrange the ten digits 0-9 to make a subtraction problem with a difference of 55,555?

GA1134

Subtraction Stars

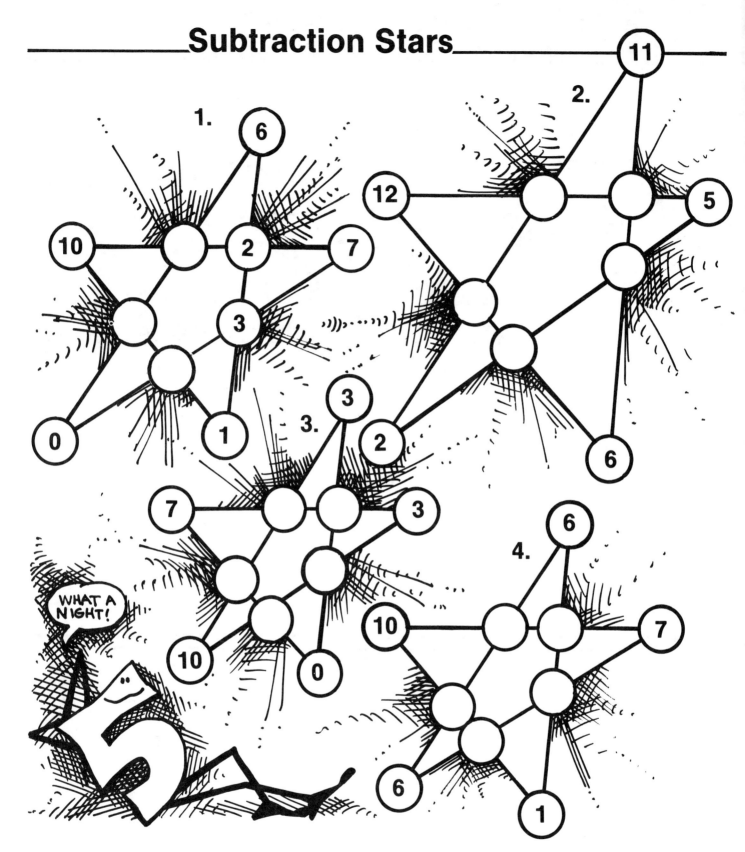

Place the five digits 1-5 in the circles so that the four numbers connected by each line is a true subtraction sentence. Example: The first two numbers of the first problem have been written in for you. 6 - 2 = 4; 4 - 3 = 1. 6, 2, 3, 1 is a true subtraction problem.

Bonus: Create a star subtraction puzzle of your own.

58

GA1134

Star Search

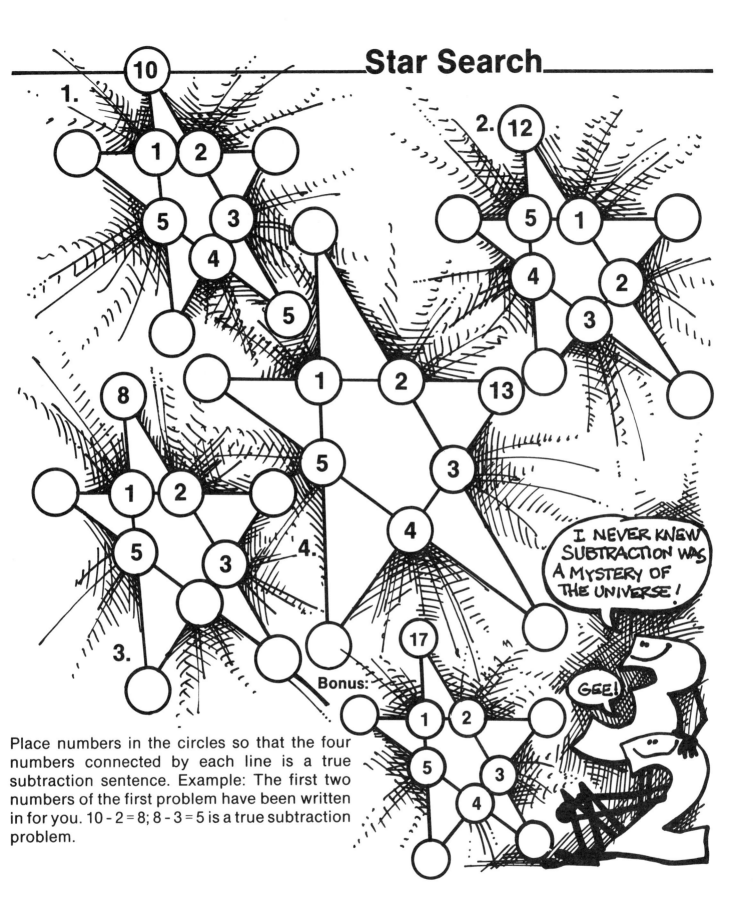

Place numbers in the circles so that the four numbers connected by each line is a true subtraction sentence. Example: The first two numbers of the first problem have been written in for you. 10 - 2 = 8; 8 - 3 = 5 is a true subtraction problem.

Bonus: Can you complete this star puzzle?

Arrange Five

Each subtraction problem found below contains the five digits 1-5. Each answer is given. You must place the correct missing numbers in each problem.

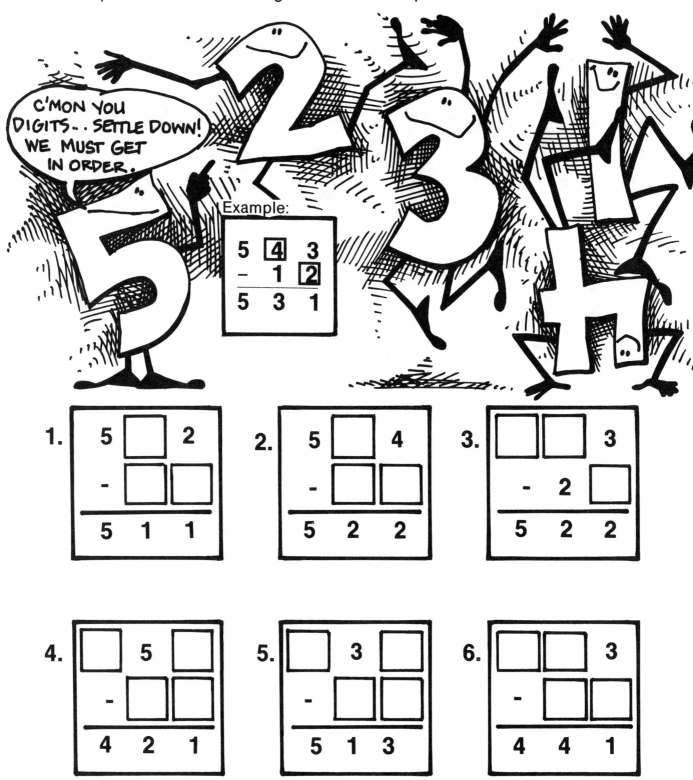

Example:

```
5 [4] 3
-  1 [2]
5  3  1
```

1.
```
5 □ 2
- □ □
5 1 1
```

2.
```
5 □ 4
- □ □
5 2 2
```

3.
```
□ □ 3
-  2 □
5 2 2
```

4.
```
□ 5 □
- □ □
4 2 1
```

5.
```
□ 3 □
- □ □
5 1 3
```

6.
```
□ □ 3
- □ □
4 4 1
```

Bonus: Can you write a subtraction problem using the five digits 1-5 with a difference of 311?

SS1134

Each subtraction problem found below contains the seven digits 1-7. Each answer is given. You must place the correct missing numbers in each problem.

Example:

7	6	5	4
−	1	2	3
7	5	3	1

1.

7	☐	☐	6
−	☐	2	☐
7	2	2	1

2.

6	☐	☐	☐
−	1	☐	☐
6	1	1	4

3.

7	☐	☐	4
−	☐	☐	☐
7	3	3	3

4.

6	☐	4	☐
−	5	☐	1
6	2	2	2

5.

5	☐	☐	3
−	☐	☐	2
5	6	2	1

6.

7	5	☐	☐
−	☐	☐	☐
7	2	4	3

Bonus: Can you create a subtraction problem with the six digits 4-9 with a difference of 333?

Sharing Candy

Three boys bought a twenty-four pound bag of candy. They want to share the candy evenly, but the only weights they have for their balance scales are five pounds and eleven pounds. What are the steps the boys should follow to get three, six-pound piles of candy?

1–1 =	5–4 =	7–0 =
12–2 =	11–0 =	12–8 =
2–2 =	8–6 =	6–1 =
4–0 =	4–2 =	9–9 =
4–3 =	12–4 =	2–1 =
8–4 =	6–5 =	8–1 =
7–2 =	11–5 =	11–9 =
7–1 =	8–5 =	5–0 =
11 –3 =	11–7 =	8–8 =
6–3 =	11–8 =	10–10 =

GA1134

12–0 =	11–6 =	5–1 =
4–4 =	12–11 =	5–2 =
12–6 =	7–6 =	11–2 =
7–7 =	3–0 =	10–9 =
7–5 =	7–4 =	10–1 =
6–2 =	5–3 =	12–7 =
12–5 =	11–4 =	10–0 =
3–1 =	9–1 =	8–7 =
6–6 =	12–3 =	12–10 =
8–0 =	10–2 =	12–12 =

GA1134

1–0 =	3–2 =	9–0 =
11–11 =	6–4 =	12–1 =
4–1 =	8–2 =	9–3 =
3–3 =	9–6 =	10–4 =
5–5 =	2–0 =	9–7 =
8–3 =	9–5 =	7–3 =
10–5 =	9–4 =	10–6 =
9–2 =	10–7 =	9–8 =
10–8 =	10–3 =	6–0 =
11–1 =	12–9 =	11–10 =

GA1134

Graphing Your Progress

Number Correct	1	2	3	4	5	6	7	8	9	10	11
90											
88											
86											
84											
82											
80											
78											
76											
74											
72											
70											
68											
66											
64											
62											
60											
58											
56											
54											
52											
50											
48											
46											
44											
42											
40											
38											
36											
34											
32											
30											
28											
26											
24											
22											
20											
18											
16											
14											
12											
10											
8											
6											
4											
2											
Test #	1	2	3	4	5	6	7	8	9	10	11

Race to the Center

Object: To land on the center space

Rules: To play this game you will need a marker and a set of dice. Begin by rolling the dice. Find the difference between the numbers indicated on the dice and advance your marker in the direction indicated by the arrows the appropriate number of spaces. (Example: If you roll a six and a four, the difference is two. You would advance your marker two spaces on the board. Keep rolling and advancing your marker on each space until you reach the center space.) This game can be played as a race between two players.

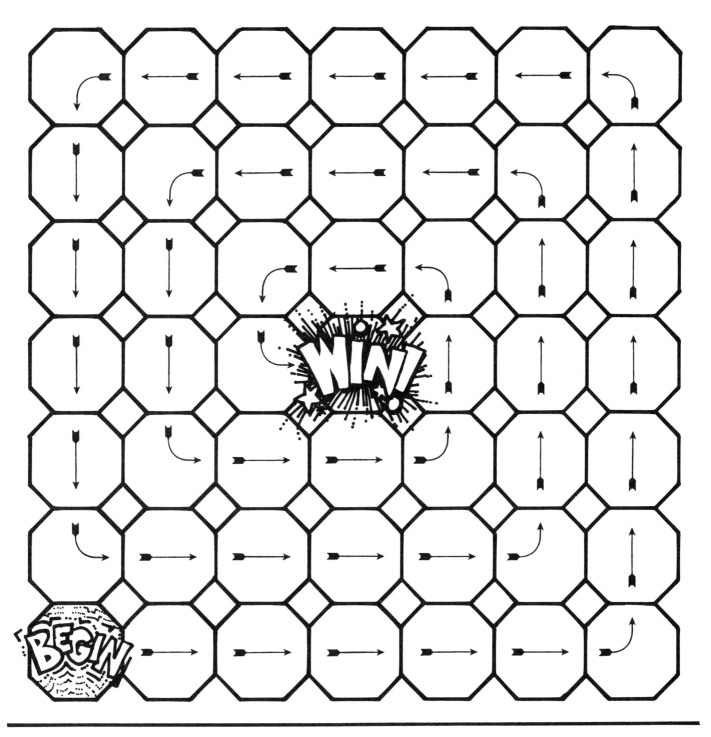

GA1134

Subtraction Bingo

Object: To roll dice with differences of the numbers on the board in less than fifty rolls of the dice

Rules: To play this game you will need fifty beans and one set of dice. Begin by rolling a pair of dice. Subtract the smaller number from the larger number. Place one bean on the space on the board that contains the difference. (Example: If you rolled a 3 and a 4, 4 - 3 = 1, so you would place your bean on the space that contains the number 1.) If you have already covered all of the spaces that contain the number one, you must throw away one bean. If you cover all of the spaces before you run out of beans, you win. If you run out of beans before you cover every number on the playing board, you lose. It's that simple. How close to a perfect game can you score?

1	2	3	4	3	2	1
2	1	0	3	0	1	2
4	5	1	FREE	1	5	4
2	1	0	3	0	1	2
1	2	3	4	3	2	1

GA1134

The Smallest

Object: To write the smallest possible number

Rules: Roll the dice three times. Each time subtract the smaller from the larger number of dots and write the difference in one of the three boxes. Then decide what is the smallest three-digit number you could have written and write it in the next box. Is your number the same? You win! Is your number larger? You lose! Try again until you have finished the page.

Example: If you rolled a five and a three, you would put the number two in one of the boxes in the first row. Roll the dice again and place the difference in one of the two remaining boxes in the first row. Repeat, this time placing the difference in the only available box left in the top row. Then decide what is the smallest possible number you could have written with your three numbers. List it in the next to the last box in the row. Did you place the three numbers in the correct order? If you did, you win. If you didn't, you lose. Repeat using each row of boxes.

	Box One	Box Two	Box Three	Smallest Number	Win or Lose?
TRY 1					
TRY 2					
TRY 3					
TRY 4					
TRY 5					
TRY 6					
TRY 7					
TRY 8					

HOW SMALL CAN YOU GO?

GA1134

Seven or Less

Object: To roll a difference of seven or less in three rolls of the dice

Rules: To play this game you will need a pair of dice and a pencil. Roll the dice and write the difference of the numbers indicated on the dice in the first box in the first row. Roll again and record the difference between the two numbers indicated on the dice. Repeat and record your third score. Next total your game score. Did you score seven or less? If you did, you win. If you scored eight or more, you lose. It's that simple. Use the second row of boxes to score your second game, etc. How many times can you beat the dice?

Roll 1	Roll 2	Roll 3	TOTAL	7 or Less?

ARE WE SEVEN OR LESS?

YES.

YES.

YES.

GA1134

Math Facts Wheel
Subtraction

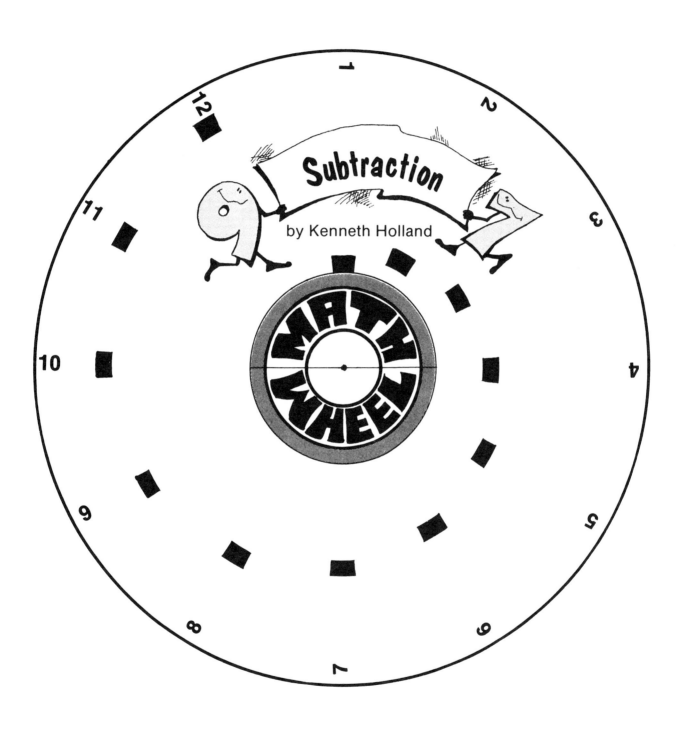

GA1134

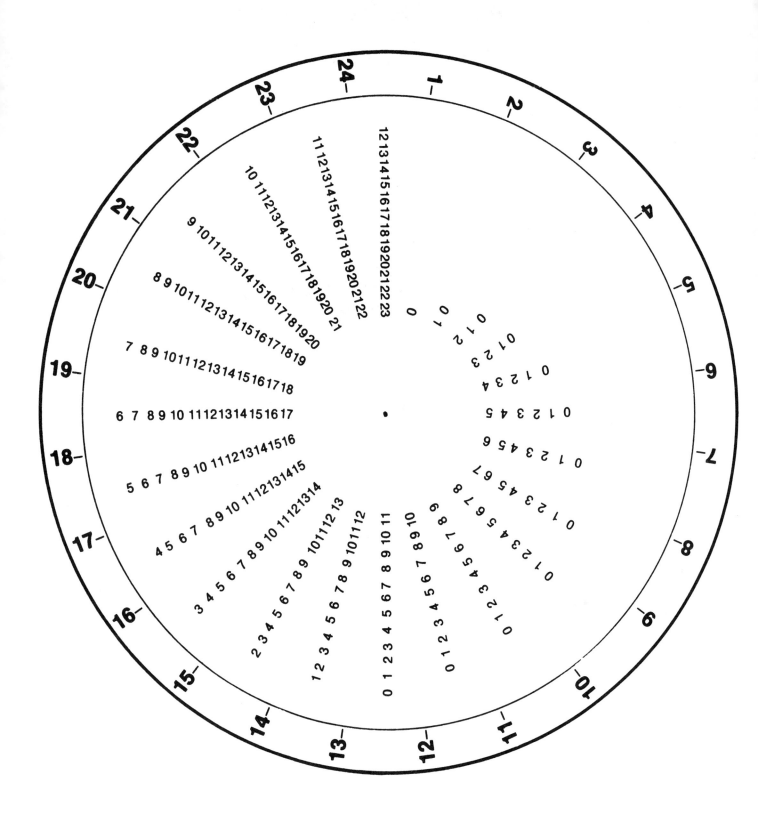

72

Answer Key

Mr. Zero, Page 1
Bonus: When you already owe him/her money.

Miss One, Page 2

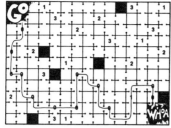

Dr. Three, Page 4
Bonus: 3 - 0, 4 - 1, 5 - 2, 6 - 3, 7 - 4, 8 - 5, 9 - 6, 10 - 7, 11 - 8, 12 - 9

Professor Four, Page 5
Answers may vary.

1. 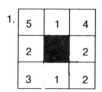 2. 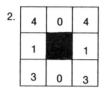 3.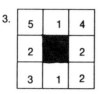

4.  5.   6.

Ms. Five, Page 6
1. 4 - 2 = 2,   6 - 3 = 3
2. 0 - 0 = 0,   5 - 2 = 3
3. 2 - 1 = 1,   7 - 3 = 4
4. 5 - 1 = 4,   8 - 2 = 6
5. 5 - 5 = 0,   7 - 5 = 2
Bonus:

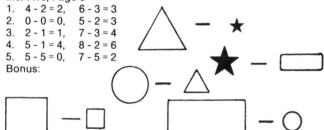

Captain Six, Page 7
1. 6 - 3 = 3,   5 - 5 = 0,   5 - 2 = 3
2. 6 - 6 = 0,   6 - 5 = 1,   6 - 1 = 5
3. 3 - 1 = 2,   4 - 4 = 0,   5 - 4 = 1
4. 6 - 2 = 4,   4 - 3 = 1,   4 - 2 = 2
5. 5 - 4 = 1,   3 - 2 = 1,   5 - 3 = 2
6. 6 - 4 = 2,   4 - 1 = 3,   6 - 3 = 3
7. 2 - 2 = 0,   1 - 1 = 0,   4 - 2 = 2

Bonus:

Here's Seven, Page 8
1. 7 - 0 = 7   11. 2 - 0 = 2   21. 7 - 5 = 2
2. 7 - 6 = 1   12. 5 - 4 = 1   22. 3 - 1 = 2
3. 1 - 0 = 1   13. 5 - 1 = 4   23. 5 - 0 = 5
4. 6 - 1 = 5   14. 6 - 4 = 2   24. 5 - 2 = 3
5. 7 - 2 = 5   15. 7 - 4 = 3   25. 6 - 0 = 6
6. 7 - 1 = 6   16. 4 - 3 = 1   26. 3 - 0 = 3
7. 6 - 2 = 4   17. 6 - 5 = 1   27. 4 - 0 = 4
8. 7 - 3 = 4   18. 4 - 1 = 3   28. 2 - 1 = 1
9. 6 - 3 = 3   19. 5 - 3 = 2
10. 3 - 2 = 1   20. 4 - 2 = 2
Bonus: 7 - 6, 6 - 5, 5 - 4, 4 - 3, 3 - 2, 2 - 1, 1 - 0

Mr. Nine, Page 10
1. 9 - 9 = 0,   9 - 2 = 7   8. 7 - 5 = 2,   8 - 3 = 5
2. 8 - 1 = 7,   6 - 3 = 3   9. 8 - 2 = 6,   7 - 4 = 3
3. 7 - 6 = 1,   3 - 3 = 0   10. 9 - 1 = 8,   8 - 5 = 3
4. 6 - 2 = 4,   6 - 4 = 2   11. 8 - 6 = 2,   7 - 3 = 4
5. 7 - 1 = 6,   9 - 6 = 3   12. 9 - 5 = 4,   7 - 2 = 5
6. 9 - 8 = 1,   9 - 7 = 2   13. 8 - 7 = 1,   6 - 6 = 0
7. 8 - 4 = 4,   5 - 5 = 0   14. 9 - 3 = 6,   6 - 5 = 1
Bonus: I – B, H – C, B – D, C – E, D – G, E – A, G – F, A – J

Subtraction Sunflowers, Page 12
2: 2 - 0, 3 - 1, 4 - 2, 5 - 3, 6 - 4, 7 - 5, 8 - 6, 9 - 7, 10 - 8, 11 - 9, 12 - 10
3: 3 - 0, 4 - 1, 5 - 2, 6 - 3, 7 - 4, 8 - 5, 9 - 6, 10 - 7, 11 - 8, 12 - 9
4: 4 - 0, 5 - 1, 6 - 2, 7 - 3, 8 - 4, 9 - 5, 10 - 6, 11 - 7, 12 - 8
5: 5 - 0, 6 - 1, 7 - 2, 8 - 3, 9 - 4, 10 - 5, 11 - 6, 12 - 7

It's Snowing Number Designs, Page 13
1. 8 – 0, 9 – 1, 10 – 2, 11 – 3, 12 – 4
2. 9 - 0, 10 - 1, 11 - 2, 12 - 3
3. 7 - 0, 8 - 1, 9 - 2, 10 - 3, 11 - 4, 12 - 5
4. 6 - 0, 7 - 1, 8 - 2, 9 - 3, 10 - 4, 11 - 5, 12 - 6
5. 5 - 0, 6 - 1, 7 - 2, 8 - 3, 9 - 4, 10 - 5, 11 - 6, 12 - 7
6. 4 - 0, 5 - 1, 6 - 2, 7 - 3, 8 - 4, 9 - 5, 10 - 6, 11 - 7, 12 - 8
Bonus: eleven

Number Pairs, Page 14
1: 1 - 0, 2 - 1, 3 - 2, 4 - 3, 5 - 4, 6 - 5, 7 - 6, 8 - 7, 9 - 8, 10 - 9, 11 - 10, 12 - 11
2: 2 - 0, 3 - 1, 4 - 2, 5 - 3, 6 - 4, 7 - 5, 8 - 6, 9 - 7, 10 - 8, 11 - 9, 12 - 10
3: 3 - 0, 4 - 1, 5 - 2, 6 - 3, 7 - 4, 8 - 5, 9 - 6, 10 - 7, 11 - 8, 12 - 9
4: 4 - 0, 5 - 1, 6 - 2, 7 - 3, 8 - 4, 9 - 5, 10 - 6, 11 - 7, 12 - 8
5: 5 - 0, 6 - 1, 7 - 2, 8 - 3, 9 - 4, 10 - 5, 11 - 6, 12 - 7
6: 6 - 0, 7 - 1, 8 - 2, 9 - 3, 10 - 4, 11 - 5, 12 - 6
7: 7 - 0, 8 - 1, 9 - 2, 10 - 3, 11 - 4, 12 - 5
8: 8 - 0, 9 - 1, 10 - 2, 11 - 3, 12 - 4
9: 9 - 0, 10 - 1, 11 - 2, 12 - 3
10: 10 - 0, 11 - 1, 12 - 2
11: 11 - 0, 12 - 1
12: 12 - 0
Bonus: thirteen

Slightly Silly Subtraction Stories, Page 15
1. 8 - 0 = 8   3. 5 - 4 = 1   5. 9 - 0 = 9
2. 7 - 6 = 1   4. 10 - 2 = 8   6. 12 - 11 = 1
Bonus: 3 games

"X" Marks the Code, Page 16
1. 8 - 5 = 3,   6 - 5 = 1,   7 - 7 = 0
2. 11 - 6 = 5,   9 - 7 = 2,   9 - 6 = 3
3. 9 - 8 = 1,   10 - 9 = 1,   12 - 10 = 2
4. 12 - 8 = 4,   11 - 7 = 4,   7 - 6 = 1
5. 11 - 5 = 6,   8 - 6 = 2,   9 - 9 = 0
6. 12 - 10 = 2,   12 - 6 = 6,   6 - 6 = 0
7. 7 - 5 = 2,   9 - 5 = 4,   12 - 7 = 5
8. 5 - 5 = 0,   12 - 5 = 7,   12 - 9 = 3
Bonus: 12 - 5 = 7 ∧ –> = <

Signing Subtraction, Page 18
1. 10 - 2 = 8   6. 2 - 0 = 2
2. 4 - 1 = 3   7. 10 - 1 = 9
3. 6 - 5 = 1   8. 10 - 8 = 2
4. 9 - 1 = 8   9. 2 - 1 = 1
5. 7 - 1 = 6   10. 3 - 2 = 1

Subtraction Bull's-Eye, Page 20

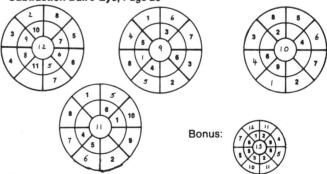

Bonus:

Down the Path of Subtraction, Page 21
2, 3, 0, 0, 1, 1

Domino Differences, Page 22
1. 7   5. 6
2. 13   6. 4
3. 9   7. 2
4. 8
Bonus: One is the most and six is the least.

GA1134

Checking the Facts, Page 23

1. 12 - 10 = 2	9. 10 - 1 = 9	17. 10 - 4 = 6
2. 11 - 9 = 2	10. 10 - 3 = 7	18. 12 - 6 = 6
3. 11 - 3 = 8	11. 11 - 4 = 7	19. 10 - 8 = 2
4. 11 - 2 = 9	12. 12 - 5 = 7	20. 11 - 5 = 6
5. 12 - 3 = 9	13. 12 - 1 = 11	21. 9 - 7 = 2
6. 11 - 1 = 10	14. 10 - 5 = 5	22. 12 - 9 = 3
7. 12 - 8 = 4	15. 11 - 8 = 3	23. 9 - 6 = 3
8. 12 - 7 = 5	16. 10 - 2 = 8	

Bonus: zero

Circle Three, Page 24

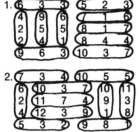

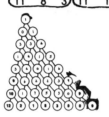

Subtraction Tower, Page 25

Subtraction Code, Page 27
Secret Message: If you learn the math facts, subtraction can be fun.

Be Mine Subtraction, Page 28
1. 8 - 2 = 6,   7 - 1 = 6,   7 - 4 = 3
2. 8 - 5 = 3,   7 - 6 = 1,   5 - 4 = 1
3. 6 - 3 = 3,   8 - 4 = 4,   8 - 3 = 5
4. 5 - 3 = 2,   2 - 1 = 1,   6 - 2 = 4
5. 8 - 0 = 8,   5 - 2 = 3,   10 - 1 = 9

Coded Subracation, Page 29
1.   11 - 10 = 1,   8 - 0 = 8,   7 - 1 = 6
2.   10 - 2 = 8,   7 - 5 = 2,   10 - 0 = 10
3.   7 - 6 = 1,   8 - 2 = 6,   7 - 4 = 3
4.   6 - 6 = 0,   11 - 9 = 2,   11 - 6 = 5
5.   11 - 4 = 7,   7 - 3 = 4,   8 - 3 = 5
6.   10 - 8 = 2,   10 - 6 = 4,   9 - 3 = 6
7.   11 - 3 = 8,   8 - 7 = 1,   11 - 7 = 4
8.   11 - 7 = 4,   11 - 5 = 6,   9 - 8 = 1
9.   10 - 5 = 5,   10 - 3 = 7,   8 - 2 = 6
10.   7 - 7 = 0,   10 - 8 = 2,   10 - 7 = 3
11.   8 - 8 = 0,   5 - 5 = 0,   8 - 1 = 7
12.   10 - 4 = 6,   11 - 2 = 9,   7 - 0 = 7
Bonus: 13 - 12 = 1

Tic-Tac Subtract, Page 30
1.   11 - 10 = 1,   11 - 9 = 2,   9 - 8 = 1
2.   9 - 9 = 0,   10 - 10 = 0,   11 - 4 = 7
3.   12 - 7 = 5,   12 - 11 = 1,   6 - 6 = 0
4.   5 - 4 = 1,   11 - 11 = 0,   12 - 4 = 8
5.   7 - 4 = 3,   6 - 5 = 1,   10 - 4 = 6
6.   12 - 6 = 6,   5 - 5 = 0,   6 - 4 = 2
7.   7 - 7 = 0,   11 - 8 = 3,   7 - 6 = 1
8.   11 - 7 = 4,   7 - 5 = 2,   12 - 10 = 2
9.   9 - 4 = 5,   12 - 5 = 7,   11 - 6 = 5
10.   12 - 9 = 3,   8 - 4 = 4,   10 - 5 = 5
Bonus: five

Search and Circle, Page 33

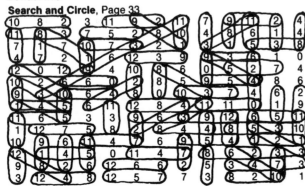

Rolling Differences, Page 34
0 = 6, 1 = 10, 2 = 8, 3 = 6, 4 = 4, 5 = 2
1, 5
Bonus: approximately zero and six, one and two

Sub Matchup, Page 35
practice
Bonus: memorize

Computer Subtraction, Page 36

1. 6	5. 12	9. 10
2. 4	6. 8	10. 8
3. 6	7. 12	11. 14
4. 14	8. 10	12. 4

Bonus: 1. 7, 2. 5, 3. 5, 4. 13, 5. 11, 6. 9, 7. 13, 8. 11, 9. 9, 10. 7, 11. 15, 12. 3

Arrow Subtraction, Page 37

1. 9 - 1 = 8	9. 8 - 2 = 6
2. 8 - 1 = 7	10. 8 - 6 = 2
3. 9 - 2 = 7	11. 5 - 4 = 1
4. 9 - 6 = 3	12. 4 - 3 = 1
5. 9 - 3 = 6	13. 7 - 6 = 1
6. 9 - 7 = 2	14. 3 - 2 = 1
7. 8 - 7 = 1	15. 2 - 1 = 1
8. 9 - 5 = 4	

Same Difference, Page 38
1. 1
2. 3
3. 5
4. 2
5. 6
6. 4
7. 8
8. 9
9. 7
10. 0

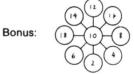

Follow the Leader, Page 39

1. 9	7. 13
2. 11	8. 4
3. 14	9. 8
4. 10	10. 7
5. 6	11. 5
6. 12	12. 15

Bonus:

1. 17	4. 18	7. 21	10. 15
2. 19	5. 14	8. 12	11. 13
3. 22	6. 20	9. 16	12. 23

Plenty of Twenty, Page 40
Answer will always be eighteen.
Bonus: Answers are always the same.

Subtraction-Go-Round, Page 41

Bonus:

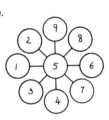

Circle Search, Page 42
1.   6 - 3 = 3,   12 - 7 = 5,   11 - 6 = 5,   10 - 5 = 5
2.   12 - 2 = 10,   11 - 7 = 4,   12 - 6 = 6,   10 - 3 = 7
3.   12 - 3 = 9,   11 - 2 = 9,   10 - 8 = 2
4.   10 - 6 = 4,   12 - 4 = 8, 12 - 3 = 9,   11 - 4 = 7
Bonus: 12 - 0 = 12, 12 - 12 = 0, 21 - 0 = 21, 21 - 21 = 0

Down the Path, Page 43

1. 12 - 7 - 2 = 3	6. 9 - 7 - 2 = 0
2. 11 - 5 - 4 = 2	7. 12 - 11 - 0 = 1
3. 10 - 6 - 1 = 3	8. 11 - 6 - 3 = 2
4. 12 - 8 - 5 = 0	9. 11 - 7 - 2 = 2
5. 10 - 8 - 1 = 1	10. 12 - 9 - 1 = 2

Cut the Cake, Page 44

GA1134

Shamrock Subtraction, Page 45
Secret Message: Because its capital is always Dublin (doubling).

Hide-and-Seek Nines, Page 46
First and last digits will always total the center number.
Bonus: Center digit is always nine.

Nine Is Divine, Page 47
Bonus: The total is always nine.

Nutty Nifty Number Nine, Page 48
Bonus: The total is always nine.

Pick Four Numbers, Page 49
Bonus: The answer will always be 10,989.

Subtraction Patterns, Page 50
1. 5, 3, 2 (subtract 1, subtract 2, repeat, etc.)
2. 11, 8, 7 (subtract 1, subtract 3, repeat, etc.)
3. 6, 6, 4 (subtract 2, subtract 0, repeat, etc.)
4. 8, 5, 3 (subtract 2, subtract 3, repeat, etc.)
5. 4, 3, 3 (subtract 6, 5, 4, 3, etc.)
6. 8, 7, 2 (subtract 5, subtract 1, repeat, etc.)
7. 16, 14, 4 (subtract 10, subtract 2, repeat, etc.)
8. 7, 3, 0 (subtract 5, subtract 4, repeat, etc.)
Bonus: 85, 79, 72 (subtract 1, 2, 3, 4, 5, etc.)

Home Run Subtraction, Page 51
1. Anne and Victor   4. 3 points
2. same difference   5. James and Victor
3. same   6. Valerie and Donna
Bonus: Donna and James

Give Me a Sign, Page 52
1.	4 - 1 = 3	12 - 11 = 1
2.	8 = 11 - 3	12 = 12 - 0
3.	3 = 11 - 8	12 - 10 = 2
4.	4 - 3 = 1	4 - 2 = 2
5.	5 - 2 = 3	11 - 7 = 4
6.	1 = 5 - 4	4 = 12 - 8
7.	5 = 6 - 1	11 - 8 = 3
8.	12 - 0 = 12	9 = 11 - 2
9.	11 = 12 - 1	0 = 4 - 4
10.	11 - 10 = 1	0 = 10 - 10
11.	2 = 11 - 9	3 = 12 - 9
12.	12 - 2 = 10	11 - 6 = 5
13.	9 = 12 - 3	11 - 1 = 10
14.	12 - 6 = 6	8 = 12 - 4
15.	5 = 12 - 7	7 = 12 - 5
16.	11 - 5 = 6	11 - 1 = 10

Two Signs, Please, Page 53
1.	69 - 5 = 64	114 - 7 = 107
2.	72 - 12 = 60	110 - 10 = 100
3.	38 - 20 = 18	189 - 45 = 144
4.	48 - 7 = 41	179 - 172 = 7
5.	97 - 14 = 83	187 - 107 = 80
6.	69 - 9 = 60	609 - 600 = 9
7.	74 - 23 = 51	147 - 47 = 100
Bonus: 1009 - 9 = 1000

Connecting Pairs, Page 54

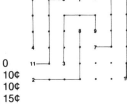

The Sweet Shop, Page 55
1. 5¢		6. 0
2. 15¢		7. 10¢
3. 5¢		8. 10¢
4. 20¢		9. 15¢
5. 0		10. 0
Bonus: Yes

Subtraction Check, Page 56
1.	6 = 7 - 1	7. 6 = 9 - 3
2.	11 = 12 - 1	8. 5 = 12 - 7
3.	9 = 12 - 3	9. 5 = 8 - 3
4.	8 = 9 - 1	10. 2 = 9 - 7
5.	7 = 8 - 1	11. 2 = 3 - 1
6.	4 = 7 - 3	12. 1 = 9 - 8

Making Special Arrangements, Page 57
1.   44 - 4 = 40
2.   666 - 66 = 600
3.   777 - 7 = 770
4.   9999 - 999 = 9000
Bonus: 98765 - 43210 = 55,555

Subtraction Stars, Page 58

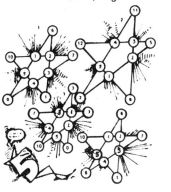

Star Search, Page 59

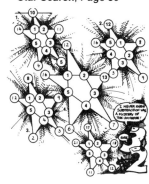

Arrange Five, Page 60
1. 542	2. 534	3. 543
− 31	− 12	− 21
511	522	522
4. 452	5. 534	6. 453
− 31	− 21	− 12
421	513	441

Bonus: 325
− 14
311

Arrange Seven, Page 61
1. 7346	2. 6257	3. 7654
− 125	− 143	− 321
7221	6114	7333
4. 6743	5. 5763	6. 7564
− 521	− 142	− 321
6222	5621	7243

Bonus: 789
− 456
333

Sharing Candy, Page 62
The answers for this puzzle will vary. Here is one way to divide the candy in three equal parts in four steps.
1. Place 11-pound weight on one side of scale and balance it with candy. Then you have an 11-pound stack and the remaining stack weighs 13 pounds.
2. Using the 5-pound balance, remove 5 pounds of candy from the 13-pound stack. That will leave an 8-pound stack of candy for the first boy.
3. Put the 11-pound stack on the opposite side of the scale with the 13-pound balance. Balance with 2 pounds of candy from the 5-pound sack, leaving a 3-pound stack of candy.
4. Place the 3-pound stack of candy with the 5-pound stack of candy to give the second boy his share. All the candy left (8 pounds) belongs to the third boy.

Pre/Post Test, Page 63
0, 1, 7
10, 11, 4
0, 2, 5
4, 2, 0
1, 8, 1
4, 1, 7
5, 6, 2
6, 3, 5
8, 4, 0
3, 3, 0

Pre/Post Test, Page 64
12, 5, 4
0, 1, 3
6, 1, 9
0, 3, 1
2, 3, 9
4, 2, 5
7, 7, 10
2, 8, 1
0, 9, 2
8, 8, 0

Pre/Post Test, Page 65
1, 1, 9
0, 2, 11
3, 6, 6
0, 3, 6
0, 2, 2
5, 4, 4
5, 5, 4
7, 3, 1
2, 7, 6
10, 3, 1

GA1134

MATHEMATICAL *Genius!*

You are great!

Unbelievable!

To: _____

For: _____

_____
Signature

_____
Date

MATH AWARD

To: _____

knows all the
subtraction facts.

_____
Signed

CON-GRAT-U-LA-TIONS!!!

You completed
_____
Math Bonus Problems

_____
To

_____
From

_____
Date

GA1134